# AUTOMATICALLY
# DRIVEN

JOHN SPARROW

First published in Great Britain in 2023 by
John Sparrow, in partnership with whitefox publishing

www.wearewhitefox.com

Copyright © John Sparrow, 2023

ISBN 978-1-915635-83-9
Also available as an eBook
ISBN 978-1-915635-84-6

John Sparrow asserts the moral right to be identified
as the author of this work.

All rights reserved. No part of this publication may be reproduced,
stored in a retrieval system or transmitted in any form or by any means,
electronic, mechanical, photocopying, recording or otherwise,
without prior written permission of the author.

While every effort has been made to trace the owners of copyright material
reproduced herein, the author would like to apologise for any omissions and
will be pleased to incorporate missing acknowledgements in any future
editions.

Designed and typeset by Typo·glyphix
Cover design by Simon Levy
Project management by whitefox

For Annette, Gillian, Evangelie and Celestine

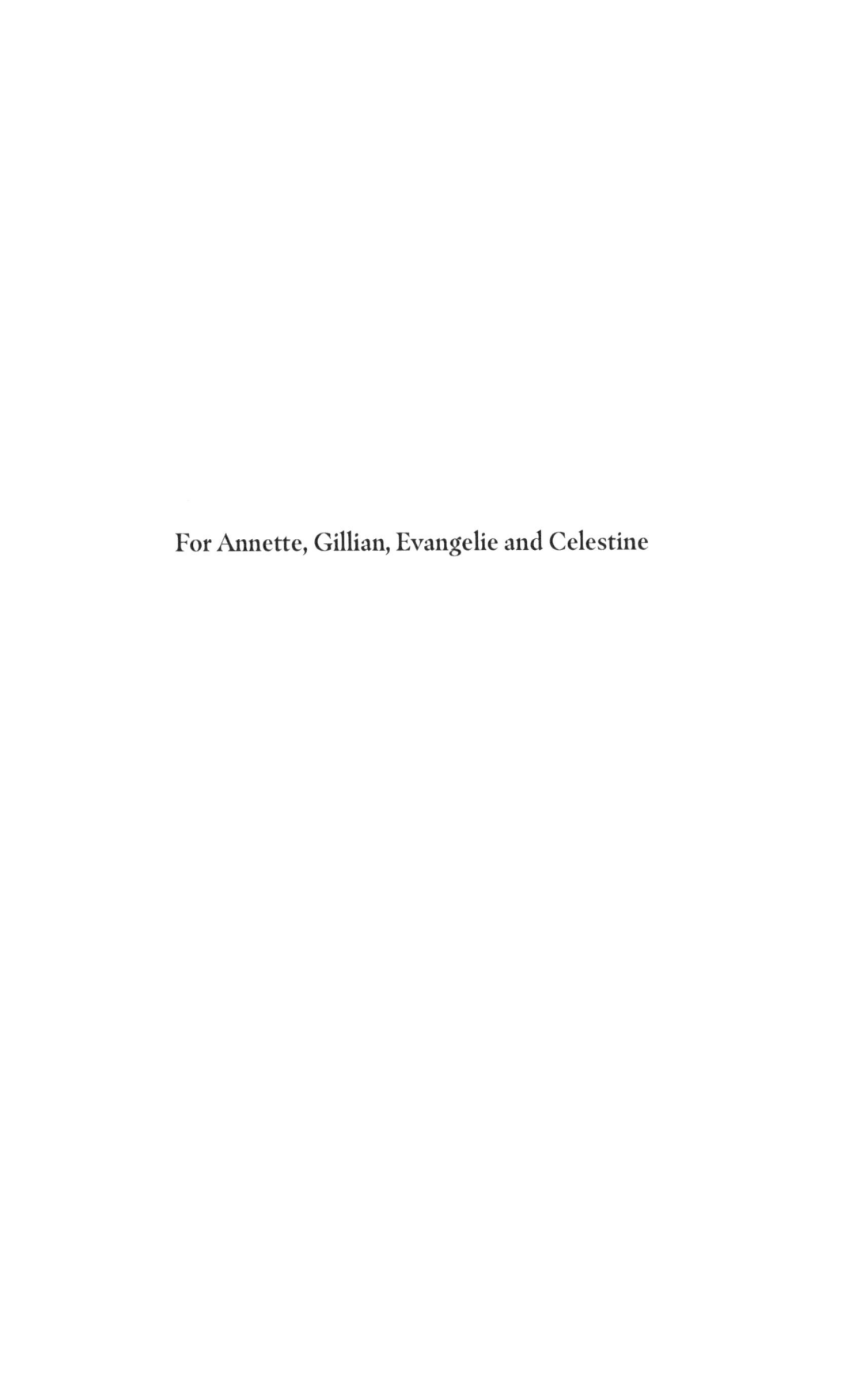

# Contents

| | |
|---|---|
| Introduction | 1 |
| Chapter 1: Car-related Life Up to 1966 | 3 |
| Chapter 2: Car-related Life at University 1966–1969 | 9 |
| Chapter 3: British Leyland 1969–1971 | 17 |
| Chapter 4: British Leyland 1972–1974 | 36 |
| Chapter 5: British Leyland 1975–1978 | 51 |
| Chapter 6: Territorial Army Exercise 1978 | 68 |
| Chapter 7: British Leyland (BL Cars) 1978–1981 | 78 |
| Chapter 8: Land Rover Ltd 1982–1984 | 96 |
| Chapter 9: Land Rover Ltd 1985–1988 | 112 |
| Chapter 10: Land Rover Ltd 1989 | 132 |
| Chapter 11: Land Rover Discovery | 139 |
| Chapter 12: Land Rover Ltd End 1989–1990 | 164 |
| Chapter 13: Range Rover North America 1990 | 178 |
| Chapter 14: Land Rover Ltd 1991 | 185 |
| Chapter 15: Land Rover Ltd 1992–1993 | 197 |
| Chapter 16: Rover Group 1993 | 222 |
| Chapter 17: BMW Group 1994–1997 | 229 |
| Chapter 18: Automotive Consultancy 1998–2000 | 252 |
| Chapter 19: Honda Motor Europe 2000–2009 | 258 |
| Chapter 20: Consultancy 2009–2019 | 290 |
| Acknowledgements | 298 |

# Introduction

**I HAVE WRITTEN THIS BOOK** for my daughter Gillian and my two granddaughters Evangelie Sinfield (short name Evie) and Celestine Sinfield (short name Lettie) so that especially for Evie and Lettie they will learn much about their grandad's life.

I wish I knew more about my father's life, especially in the years 1940–1945 when he was with the British Army fighting in North Africa. Although I have his full army records there are no details of what he did other than training and time in the army hospital. When you are young you rarely ask the questions to the elders in your family that you wish you had asked when you are much older. By then it is usually too late because when they are gone the answers to your questions have almost certainly gone with them.

I think it is unusual for a person to want to write a book about himself or herself, but also, and more importantly, to have the desire, commitment and time to do it. I knew I wanted to write a book for the younger members of my family because if I didn't almost all of my memories and experiences would disappear with me when I have gone. My daughter Gillian said to me, 'Write it.' So I have.

This book is not a comprehensive life story but more a compilation of things that happened to me, things I remember and things I did or experienced. I wanted to document some specific memories, experiences and events of my time and career of fifty years in the car industry. I was fortunate to join British Leyland direct from university and to have progressed through management positions to some senior Executive and Director roles in Land Rover, Rover Group, BMW Group and Honda Motor Europe. I don't think many people alive today have worked at these senior levels for wholly owned British, German and Japanese motor car manufacturers.

While most of this book is based on my personal memories, I have been helped in describing details and dates on some subjects and specific events by various documents and the many pocket diaries that I have kept over the years. I suppose I thought they could be useful one day and they have definitely helped me write about specific events. If anyone is thinking of writing a book about their life then I strongly suggest keeping a daily diary as it will prove to be very valuable.

The content of the book is broadly in chronological order but with the passage of time I cannot guarantee total accuracy on some dates, events or names. Where possible, and for the record, I have included the names of the many colleagues and car dealers with whom I worked in my long car industry career.

I decided to stop writing in September 2019 to coincide with my reaching fifty years in the car industry. This seemed a logical point to finish my book. I was lucky in many ways, not least because I was inwardly and automatically driven to have a successful career in the industry, something I achieved and enjoyed in doing so.

I hope everyone who reads this book will find some of it interesting and in places maybe a little educational.

JOHN SPARROW

# Chapter 1
# Car-related Life Up to 1966

**I WAS BORN ON 14TH SEPTEMBER 1947** at the Osterhills Hospital in St Albans, Hertfordshire, England. On my birth certificate I was named John Alexander Sparrow. At the time of my birth my father and mother were living at the home of my father's parents at 93 Necton Road, Wheathampstead, a village a few miles north of St Albans.

Until I went to the infants' school in Wheathampstead at the age of five, I am fairly sure there was little for me to do or to occupy myself with each day. However, I do not think I was ever bored so I must have found things to do. You have to remember in the early 1950s we had no television entertainment and I don't remember having any books to read so I played most of the time with some small metal dinky cars and motor racing cars. I am sure my lifelong interest in motor cars started then.

Around the age of four I received a small metal pedal car as a Christmas present. I remember we went on the train from Wheathampstead to Luton to buy it. The train station at Wheathampstead and the railway line from Luton to Wheathampstead to Welwyn Garden City closed in 1965 under a major closure of mainly rural railway lines in the country. However in the 1950s it was an important means of transport as very few people owned cars.

As a very young boy cars had become my main interest. My father used to come home for lunch, or dinner as it was called then, and he would have to push me round the garden in my pedal car because I wanted to go faster than I could by pedalling myself. I asked him to create two doors in the metal panelling, and a number plate with my initials on the front and back of the car, which he did. I also asked him to fit some lights on the car, which, although it was a rather extravagant request, he did as well.

I suppose you could regard these requests to upgrade the specification of my pedal car as my first attempts at vehicle product development, a subject which I would become involved in during my career in the car industry.

There were around eighty houses in the U-shaped Necton Road. However, down the eastern side of the road I only remember one person with a car. Almost opposite our home at 54 Necton Road was number 69 and I could look out of our front-room window and there often on the drive was a rare and expensive Alvis sportscar owned by a Mr Jack Fegan. I don't know what he did for a living but he must have been fairly wealthy because his Alvis was a very rare and expensive car at that time. I was very envious, because to me, even at that young age, it was an exceptionally elegant and desirable motor car.

I suppose that's the sort of thing you remember as a four-year-old. It made a great impression on me and I wanted our family to have a motor car. We were not alone as very few people in the UK had a car at that time. At first those that could afford it had to make do with pre-war used cars, as rationing and the drive for exports by the government after the war ended in 1945 restricted the availability of new cars on the UK home market.

Most cars on the road at the time were made in Britain and mainly comprised Austin and Riley cars made at Longbridge, Birmingham, Morris and Wolseley cars made at Cowley, Oxford, Fords made at Dagenham, Essex and Vauxhalls made at Luton, Bedfordshire. There were relatively very few imported cars, these mainly coming from France (such as Renault), Italy (mainly Fiats) and Germany (mainly Volkswagen and Mercedes).

One of the most popular cars at the time was the Morris Minor. The Morris Minor was built at Cowley in Oxford from 1948 until 1972 and was Britain's first car to sell over a million. A twelve-year-old

1956 Morris Minor would become my first car. More on that later. Once rationing was finally over there was a relative boom in car ownership. The car and the booming British car industry were symbols of the improving prosperity in the country.

Around 1955 my father got his first motor car, a Vauxhall Wyvern saloon, with the registration number OUR342. That number plate alone would be worth several thousand pounds if we still owned it today. Up until then he travelled to and from his work as a builder, and to his local council meetings, on a relatively small motor bike, called a New Hudson. New Hudson motorbikes were a subsidiary of BSA but they stopped making them in 1957.

The arrival of my father's Vauxhall motor car made a great difference to our life as I recall we often went out on a Sunday afternoon for a drive around the villages of north Hertfordshire. We take travel by car for granted today, but having a car then was life-changing for my family, and very exciting for me, not yet eight years old.

I can't remember anything about the eleven-plus examination. However, I must have done well in the exam because I passed and won a place at the St Albans County Grammar School for Boys.

Everyone in the family was very proud of what I had achieved as I was the first person from the Sparrow family to go to a grammar school. Although I didn't know it at the time passing my eleven-plus exam and going to the St Albans Grammar School was one of the most important events in my life as it set me on course for going to university and eventually a successful career in the motor industry.

In 1960 my father changed his Vauxhall Wyvern for a new Morris Oxford Estate. The car was bright green and had the registration number 9005 AR. I was aware that this was a vehicle purchased as a company car through Smith Brothers, where my father was a Director. This Morris was a step up from the Vauxhall.

In 1962, at the start of the fourth year at St Albans Grammar School, some important decisions had to be made about which subjects to take. These decisions would determine whether you followed a 'science' route or what the school called a 'modern' route. I had already decided I wanted to go into business in the motor industry, a dream at that point – I didn't know what that meant in practice. Accordingly, I chose to take the 'modern' subjects of languages and geography which I was quite good at and enjoyed, rather than physics, chemistry and biology which I did not enjoy and which maybe not surprisingly were my weakest subjects.

I had already decided that providing I passed the O-levels, I would take Geography and History at GCE A-Level as these were only a few of the subjects I could now study with Economics and Public Affairs. This was a fairly new A-Level on the syllabus and I would be taught by an excellent Welsh Schoolmaster called Gavin Thomas, who would prove to be very influential not only in the next part of my education at St Albans Grammar School but also on my decision on which university I would apply to go to.

In addition to what I would call normal school homework over the Christmas holiday at the end of 1964 I was required to research and then write a study on a local industry. I had the possibility of doing Helmets or Murphy Chemicals, both relatively small companies operating in Wheathampstead at that time, but neither of these appealed. In St Albans the most well-known company was L Rose & Co Ltd, which made lime juice cordial and lime marmalade, but that company definitely did not appeal to me.

In the end I received approval to study a local industry that did appeal to me, namely Vauxhall Motors Ltd, based at Luton, Bedfordshire. I needed to get approval because I wasn't sure whether it could be regarded as local, even though it was only just over 7 miles from where I lived in Wheathampstead.

While doing the research and getting to know about Vauxhall as a manufacturer of motor cars I became even more interested in the possibility of working in the car industry, hopefully after going to university. There is no doubt in my mind that being required to do research on Vauxhall Motors and then write a fairly detailed report within a few weeks greatly helped me obtain an appreciation of the motor industry business for the first time.

In March 1965 a group of us went on an Economics–Geography field course for a week to South Wales. The group comprised those of us studying Economics and Geography GCE A-Levels and we were led by Mr G. Thomas, Economics Master, and Mr E. Alexander, Geography Master.

I can vividly remember being in a small group that went down the coal mine at Bargoed, an eye-opening experience into the very hard and dirty working life of the coal miners there. At the end of the week's field course, Mr Thomas had arranged that, while we were in the Swansea area, our coach would make a short visit to University College Swansea, a part of the University of Wales at that time.

The main message Mr Thomas gave to us was this is the best university in the country if we wanted to study for a degree in Industrial Economics. He told us that he had studied there and suggested those of us who were planning to study for an Economics degree should seriously consider doing the same. The university had a great location on the Mumbles Road overlooking Swansea Bay and with my Economics Master's recommendation, this would be seriously taken into account a year or so later when I had to decide to which university to apply.

Sometime into early 1966 we were asked to start thinking about which universities we wished to apply to. I wanted to study Economics at university, and so with the recommendation of Mr Thomas, I decided Swansea would be my first choice.

Shortly after submitting my application, I received a written communication confirming an offer of a place for me at the University of Wales, University College Swansea from September 1966, on condition I obtained three GCE A-Levels at grade C or better or two GCE A-Levels at grade B or better. At the time, GCE A-Level grades were awarded on a normal distribution curve basis so only a few people received A grades, more received B grades, even more C grades, then reducing numbers of D and E grades.

When I received my exam results, I was delighted and relieved to find I had obtained grades more than good enough to go to the University of Wales at Swansea. I could now start to think about preparing for a life away from home.

I was pleased to find out that John Bromley and Brian Wilkinson from my school in St Albans would be going also to Swansea to study for a similar BSc Econ degree, and that Stephen Jones and Richard Jones (no relations) would also be going to Swansea to study other subjects. Mr Gavin Thomas had definitely had some positive effect and influence on the five of us.

Chapter 2

# Car-related Life at University 1966–1969

**THERE WAS A POINT DURING MY SECOND YEAR** at university when I decided I wanted a car and so it was time to have some driving lessons and hopefully take and pass the driving test. It was clear that having a car at university would make travel easier and more convenient in Swansea, as well as giving me much more freedom and the opportunity to go where and when I wanted. In addition, it would make life much better for the long 200-mile journeys to university from home at the start of each term and the return journey at the end of each term. It should be borne in mind that relatively few students owned a car at that time.

Some time during the summer of 1967 I had received a provisional driving licence and soon after my dad suggested taking me out for a lesson. I put some L-plates on my dad's Austin 1800 and one afternoon went out with him on a so-called driving lesson. He didn't give me much instruction or advice before we set off from home in Wheathampstead, so it was more a question of 'see how it goes'!

For some reason he told me to go down Dyke Lane towards St Albans. Dyke Lane was a road just wide enough for one vehicle and with very few places in its mile length for two to pass. Inevitably a car was coming the other way. Rather than gesturing for the other driver to reverse, my father told me to reverse, which was not easy to do in this narrow winding road, barely five minutes into my first time of driving a car, which was also quite wide. Somehow he expected me to reverse perfectly the first time and was not happy and far from helpful when I was not doing it right. On this occasion, sympathetic and supportive as a teacher my dad was not. Based on this limited experience, I didn't think he was a good driving instructor, so I decided that would be the last time I would go out with him.

A few months later back in Swansea I booked some lessons with a driving school and was pleased to have a good instructor. I told him I had not driven before. I also decided I wasn't going to tell my dad or the rest of the family that I was taking driving lessons. The one-hour lesson was spent driving an Austin 1100 in and around Swansea. I particularly remember we were driving for most of the time in the Townhill and Uplands areas, which meant – as the names imply – a fair amount of uphill and downhill driving content. I remember one of the hardest parts of the driving lessons was staying within the 30-mile-an-hour speed limit on some of the long downhill stretches.

Before long, my instructor was taking me on a route that I was told would probably cover most of the driving test route, and the lessons now included reversing around corners, three-point turns, hill starts and emergency stops. After five or six lessons I was told I was progressing very well and a date for my test was booked.

On the day of the driving test in March 1968 I had what I hoped would be my final lesson. I think it was my tenth, and it greatly helped that it was a fine dry day. My instructor told me that the driving test inspector I would be getting that day had the habit of saying something like, 'When I hit the dashboard with my newspaper I want you to do an emergency stop.' I would remember that!

I felt the test was going well and I didn't think I had made any mistakes. We were broadly following the route I knew quite well so that helped my confidence. Then near the end of the test we travelled along the road where I was told the emergency stop could take place. Shortly after I could see out of the corner of my eye that the inspector was lifting his newspaper, and at the very moment he hit the dashboard with it I jammed my two feet on to the clutch and brake so hard that his body lurched forward. For an instant I thought I was going to send him through the windscreen. As far as I was concerned, I had performed a perfectly controlled emergency stop. His face had no expression and he said nothing so at that moment I didn't know if he agreed with me. However, I was very pleased the test inspector was wearing a seatbelt!

At the end of the driving test the test inspector again had no expression on his face and I couldn't tell what the outcome would be. After a moment or two he just said, without looking at me, a few words I shall always remember, 'I think I will give you a pass.' I think I just said, 'Thank you.' I then received a few comments from him about the procedure I had to follow to obtain my full driving licence, which I did without much delay. I was absolutely delighted to have passed at my first attempt and thanked my driving instructor for not only teaching me how to drive properly and safely but also in preparing me for the test itself.

I would be going home shortly after for the Easter holiday and I would have great pleasure telling my dad in particular and the rest of the family I had passed my driving test. In addition, I could now consider buying my first car during the few weeks on holiday and returning to Swansea for the summer term with it.

I was pleased to tell everyone in the family I had passed my driving test. In view of our single unsuccessful driving lesson together I think my dad was both surprised and pleased. I told him I would hope to buy a cheap second-hand car for around £100 (about £1,700 in 2020 money) during the holiday so I could go back to Swansea with it. He said he would be pleased to help me financially with the purchase and I was very grateful to him for that help.

I started looking through the motoring section of the *Herts Advertiser* where cars were advertised for sale. This was the local weekly newspaper and from what I remember there were not too many used cars for sale at the price I had in mind. The used cars that were advertised by franchised car dealers were too expensive for me and in the end there were just a couple of cars being advertised privately by local owners that I could afford and consider.

The car I eventually bought for £100 was a Morris Minor, registration number MAP204, first registered in 1956, so it was twelve years old. It was the model first introduced with a one-piece front windscreen, and the car I bought was bright green, with four doors and leather seats. It also had retractable red semaphores or 'trafficators', rather than flashing lights, mounted high up behind the front doors and which swung out horizontally, if you were lucky. These were still common on vehicles built in the 1950s.

Cars registered after the end of 1958 required flashing direction lights, which didn't include my Morris. However, with the help of my friend Mike Good and the garage he worked at I had flashing direction indicators fitted to the front and rear of the Morris. For safety reasons it seemed the right thing to do. My father very generously paid for the Private Car Third Party Fire and Theft car insurance annual premium of £29.

In 1965 a requirement for seatbelt anchorage points was introduced in the UK, followed in 1968 by the requirement to have three-point belts fitted in the front outboard positions on all new cars and all existing cars back to 1965. For safety reasons again, even though there was no requirement to have them on my 1956 Morris Minor, I had front seatbelts fitted.

In due course I set out on the trip to Swansea. Six hours and around 200 miles later I arrived. I had not driven my Morris Minor or any car very far before, so this journey was a long one in every sense. But I was happy the car and I got there safely.

I had some bad luck on my journey from home to Swansea in the winter/spring term of 1969. My Morris Minor suffered a breakdown on the A4 as I was entering the town of Marlborough in Wiltshire. I remember coasting down a slight decline into the town and fortunately had enough momentum to reach a small vehicle service garage. The water pump had broken so there was no way I could continue. The garage said they could fix it but not that day as they did not have the parts. So, I thought, how do I get to Swansea? I had an idea.

I was aware that my flatmates John and Brian were also travelling the same route that day. I didn't know where they were and there were no mobile phones then to call them. I asked the garage if they would let me phone the police to see if they could stop either John in his Ford Thames van or Brian in his Minivan and ask one of them to come and pick me up from the garage in Marlborough.

John must have had a real shock when he was stopped by a police car further along the A4 and was told he had a friend who needed his help back in Marlborough. I was of course very grateful to John for coming to pick me up, but also to the police for their help in resolving what for me was an emergency. However, I am fairly sure the police in Wiltshire today would not be willing to help a broken-down motorist as they did for me.

A week later I made a train journey from Swansea to Swindon, the latter being the nearest railway station to Marlborough. From there I was fortunate to hitch a lift quite quickly direct to Marlborough. I think the bad luck of breaking down was compensated by the good luck I received in getting immediate help from the local police and the good fortune to have my flatmate not far away and willing to come and get me.

Not long afterwards some more bad luck. My Morris Minor was stolen from where I usually parked it in a side road near our flat in Glanmor Road. Fortunately I had some good luck when the police soon found it not far away, and I got it back undamaged.

During our last year at university we were informed about the annual 'milk round' – the term commonly used in the UK to describe the procedure whereby companies visit universities each year, in order to advertise their career opportunities and recruit students.

At the time I was really only interested in joining a motor vehicle manufacturer. I had done a study of Vauxhall Motors at school and the motor industry was part of my Special Industries subject in my Economics degree. I was very clear in my own mind about the companies I wished to see when they came to Swansea. It was a shortlist of four.

The motor industry companies I applied to see were British Leyland (BL), Ford, Vauxhall and Rootes. The latter made Hillmans in Coventry.

When the companies came to Swansea it was clear the short meetings were first interviews. I recall the one with BL started with the question, 'Why do you want to join British Leyland?' We were in open-topped booths in a big room and I could hear other students struggling to give a positive answer. The BL management made it clear they were looking to recruit university graduates who could contribute to the success of the recently formed company, and who in time would become the future leaders of the company. That was a positive message that greatly appealed to me. I was pleased to be offered a second interview by BL and Ford but not Vauxhall or Rootes. I was told later that Vauxhall was only recruiting two university graduates that year so not getting a second interview with them was understandable. Rootes was also recruiting only a very few graduates.

The second interview with BL was held at High Hall, one of the students' residential buildings at the University of Birmingham in Edgbaston, and I drove there in my Morris Minor. I say interview, but it turned out to be a very comprehensive two-day assessment. This included an individual interview as well as some intelligence tests and group exercises. On the first evening we had dinner and afterwards we were told to relax. We were bought a few drinks and some of us played bar billiards. However, I was conscious we were being 'observed' as Richard Wright, a university graduate recently recruited by BL, was there with us. I was sure he would be making his own assessment of us during the two days and would certainly be asked to convey his opinion of each of us.

The second interview with Ford was not quite what I expected. I was interviewed by a group of four managers, but what I remember most was that I was continually interrupted before I had chance to fully answer the questions they asked. I was aware this was probably deliberate to see how I reacted but I didn't feel any rapport with the interviewers, which I had done with the BL management. In addition, there was no positive message about the benefits of joining the company and the career opportunities there.

It was not too long before I received a letter from BL offering me a position as a graduate trainee in their Austin-Morris Division at Longbridge, Birmingham, in September 1969 on condition I achieved a second-class university degree. I was more than happy to have the chance to start my career where I wanted! Soon the final degree examinations came round. I can't say I was as confident as I should have been. In addition, there was the extra anxiety of knowing my planned career in the motor industry depended on getting a second-class degree. All eight of the exams were of three hours' duration and were held in a short period of ten days. When I walked out of the final exam which finished on 12th June 1969, I said to myself I never want to do another academic examination again! Before long the results were confirmed and I had achieved the second-class honours degree I needed and I was happy with that. It meant I could put BSc Econ (Hons) after my name, and take the job at BL.

My time at university in Swansea had been a great experience. Living away from home had made a big impact on me because my lifestyle changed totally. I lived in a small hotel in my first year and then in a flat for two years with three good friends. I became actively involved in the Students' Union. I played a number of sports. I enjoyed a good social life. I know I didn't work as hard as I should have. However, having reflected on my time at university I think I achieved a reasonable balance of all these things. The next big step for me was to start my working life and career in the Austin-Morris Division of BL.

More recently Swansea University completed an oral history project entitled 'Voices of Swansea University, 1920–2020: An Oral History'. The aim of the project was to record the memories and experiences of individuals who studied and/or worked at Swansea University between 1920 and 2020. I was invited to participate and in February 2017 I was interviewed at my home by Dr Sam Blaxland, Postdoctoral Research Fellow at Swansea University. For a couple of hours, I answered questions on all aspects of my life and time at Swansea University between October 1966 and July 1969. This was recorded by Sam and together with various documents I had kept from my time at Swansea, these have been deposited in the Richard Burton Archives in the Swansea University library. Sam Blaxland's book, *Swansea University: Campus and Community in a Post-War World*, has now been published.

Chapter 3

# British Leyland 1969-1971

**WHAT THE **** HAVE I COME TO?** Those were the words I said silently to myself on entering the large open-plan Austin-Morris Division Sales Block at Longbridge, Birmingham, on my first day at work on the Monday morning of 1st September 1969.

A short while before I had received a letter informing me that I had been booked into the Red Lion Inn, in Bromsgrove, Worcestershire, for the Sunday night before, for one week. I would have to find more permanent accommodation myself, and quickly.

In preparation for another period of time away from my family and my real home I prepared my Morris Minor for the journey to Bromsgrove and packed my clothes and other belongings. These included one dark suit and a few formal shirts and ties. The journey would be around 100 miles and at my Morris's cruising speed of 50 miles an hour the journey would take just over two hours.

On arriving at the Red Lion, I was pleased to learn another BL graduate trainee, as we were to be called, was also booked in for the week and this person was John King, who had driven from Scotland in his Mini. He too was going to work in the Sales Division at Longbridge and so early on the Monday morning we drove the 7 miles up the A38 to the factory.

We entered the factory at the main Q gate entrance in Low Hill Lane, parked our cars and made our way to the main reception in front of the Exhibition Hall. Eventually we were taken to the Personnel department where I recall receiving a brief welcome and introduction to the company. I will always remember what happened shortly after this.

Fred Herlihy was the Graduate Training Manager. He said he would take me to the manager of the department I would be working in and we entered the Sales Block, as it was known.

I distinctly remember the loud noise made by the clatter of the many typewriters, the cigarette smoke that formed a thin but very visible fog inside the building and the piles of paper stacked high on top of the continuous lines of filing cabinets which seemed to define specific working areas for the staff. I was taken aback and shocked by what I saw and heard, hence the strong words I spoke silently to myself.

Eventually we reached the place where I was going to work and I was introduced to Ivor Evans, Department Manager, within a Market Representation department headed by Bert Lawrence. At this point Fred Herlihy left me with the words, 'I will keep a careful eye on you.' Much later in the book I make reference to these words.

I was soon joined in this department by some other graduate trainees. We were gradually introduced to other members of the department including Arthur Challenor, who was responsible for all the distributor and dealer legal agreements, and Arthur Willets (known to us as 'Arthur the Maps'), who looked after and provided all the maps of the UK we would need in due course. We also met Des Haydon, Richard Wright and Bill Cashman, who were supervisors in the department.

I had been told I was joining the Business Management team with a view to supporting the introduction of a BL company-based management accounting system into the Austin Morris dealer network in the UK. However, within the first week I experienced my first of the very many organizational changes I would be involved in during my career, although this first one was very mild compared to some that would come later. I would not be working on accounting systems after all but instead on a major UK franchising strategy which Bert Lawrence's Market Representation department would be managing.

I quickly found out about the organizational pyramid that existed in the company and which I would need to climb if I was to reach top management level in the future. I was at the bottom level of the monthly paid staff as a graduate trainee, and I would receive £1,125 per annum in monthly sums. The role of graduate trainee would last for six months, at the end of which, all being well, I would be 'promoted' to the position of 'Analyst' with job grade 7.

Above Analyst was a Coordinator at grade 8, then Supervisor grades 9 and 10, then the first rung of management which had seven grades rising from 25 to 31. Above that were so-called ungraded Director positions, which I later found out had grades which started at C then B then A.

In addition, within the organization and below the monthly paid staff there were many hourly paid grades covering secretarial, administration and the various manual and production jobs in the factory.

I found it very hard to believe there could be so many layers of staff and management in an efficient business, as well as the seemingly very large numbers of staff employed there.

It was indeed a steep mountain of grades and there was also a factor called seniority rather than ability that could get in the way of career advancement in the company. This was explained to me in simple terms that meant I might have to wait a long time for people above me in the organization to be promoted, retire or die before I got the chance to move up the ladder.

Many graduate trainees left the company not long after joining because they couldn't see a path of promotion or of career development. Maybe some of them weren't suited to the motor industry, or specifically to BL. However, there were some of us who would spend their whole working life in the motor industry.

Some of us did climb the ladder to be Directors. Other graduates who joined BL the same day as me included David Bower, who went on to become Main Board Personnel Director of the Rover Group, and Peter Johnson, who went on to become Sales and Marketing Director in the Rover Group and later Group CEO of Inchcape plc, a

British multinational automotive distribution, retail and services company.

The Sales Block was also structured physically with due regard to grade seniority. One's actual workplace was based on one's grade or position in the department, and that within the overall organization. Almost everyone below management grade worked at their desk in their department's allocated space in the open-plan area of the Sales Block. Around the sides of the Sales Block were small 'glass box' offices for management, with their secretaries positioned at a desk immediately outside. Filing cabinets then provided physical barriers between one department and another. A corridor separated the vast open-plan office area with its ring of management offices from larger self-contained offices which were occupied by Directors and top management.

We were informed of the canteen dining arrangements. As with almost everything else, where one could have lunch depended on one's job grade in the hierarchy. I also learned that once you reached the lofty level of Director you even had a personal key to a Directors' toilet. That was the way the company operated in 1969. Where you sat and worked, where you had lunch and even where you went to the toilet depended on your job and its grade!

BL, within which Austin Morris was one car division, had recruited around sixty university graduates in 1969 in what was regarded as the company's first large-scale intake of graduates. I had already been told when interviewed that BL recognized it needed to recruit talented young people across the whole business including sales, finance, personnel, purchasing, engineering and manufacturing, not only to support the short-term business but also to provide the future leaders of the company. That particular message had appealed to me and greatly influenced my decision to apply to join BL.

Encouraged by Harold Wilson's Labour government of the time, BL had been formed in 1968 as the British Leyland Motor Corporation Ltd

(BLMC), following the merger of Leyland Motors Corporation and British Motor Holdings. The enlarged company included Austin, Morris, Riley, Wolseley, MG, Rover, Triumph, Jaguar, Land Rover as well as commercial vehicles, trucks and buses. It employed around 130,000 in the UK and up to 200,000 worldwide. BL at that time was reputedly the fifth-largest producer of motor vehicles in the world, producing around a million vehicles a year.

In the first week or so we were all gathered together to hear senior BL Directors give speeches on the objectives of BL and the important roles we were expected to play. However, what I remember more distinctly at the time was a meeting in the Exhibition Hall at Longbridge, to which we graduate trainees were summoned to attend. Here trade union officials gave us a talk on the role and importance of the trade unions in the company and within their message there was a strong indication that if the graduate trainees did not join a union we could find ourselves without jobs.

If the way the message had been conveyed had been different and less threatening, I suppose I and some others may have joined, but in the end the veiled threat backfired. I did not join and I think most of the graduates were also not persuaded. And we kept our jobs!

The trade union influence was also very noticeable within our department. The departmental secretary had some trade union position of importance because she had a thermometer on her desk to measure the temperature in the office and I was told if it was too cold, or too hot, then it could lead to work stopping.

In the first week I also noticed that when the bell started ringing at the end of the working day at ten to five there was literally a human stampede to the exit doors. The aforementioned secretary did the same, and I also observed that if the bell rang while she was typing then her typing usually stopped in mid-sentence, and the typewriter cover came down. To me this seemed unbelievable, but I was soon to learn the attitude of many working at Longbridge was to do as little as possible within the official working hours and nothing beyond that. Undoubtedly this practice must have had a serious adverse effect on productivity in the business.

During my first week at work I sounded out many of my new graduate friends about accommodation. As a result of these conversations I arranged to rent a room in a terraced house in the Kings Heath area of Birmingham where Nick Herd of the Sales Finance department was living.

Before going into detail about my job I will give some very brief details of the UK business structure in September 1969. Austin, Morris, Riley, Wolseley and MG brands each had their own distribution and dealer networks. Austin and Riley franchises tended to be held by the same dealers as both of these brands' products were made at Longbridge, Birmingham. Meanwhile Morris, Wolseley and MG franchises tended to be operated by the same dealers, as their products were made at Cowley, Oxford.

Many vehicles were badge-engineered and sold through the same dealer networks, but because they were very similar products they were often competing against each other. Badge engineering for BL was a cost-effective way of applying a different badge to an existing vehicle and then selling the badged version as a distinct product. However, this was all considered to produce inefficient distribution and too much intra-brand competition. Examples of badge-engineered vehicles within the Austin Morris model range are as follows:

- Austin Mini/Morris Mini/Riley Elf/Wolseley Hornet
- Austin-Healey Sprite/MG Midget
- Austin 1100/Austin 1300/Morris 1100/Morris 1300/MG 1100/Riley Kestrel/Riley 1300/Vanden Plas Princess/Wolseley 1100

The new Austin Morris sales organization was headed by Filmer M. Paradise, an American who had joined the British Motor Corporation

(BMC) in 1967 from Ford of Italy, and who was now Director of Sales. Reporting to him were the following: Bernard Bates, Director Home Sales; Michael Trodd, Director Export Sales; Tony Aston, Financial Controller; Michael Heelas, Vehicle Sales Manager; Henry Jelinek, Programming and Distribution Manager; and Doug Pittaway, Organization and Administration Manager.

Working with Bernard Bates was Bert Lawrence, who as previously indicated headed up the Market Representation department I was working in. Filmer Paradise was interviewed in March 1969 and from the resulting press article I have found the following words which neatly summed up Bert's role and objective.

> *Mr Lawrence has been given the task of attempting to reorganize Austin Morris's 5,000 retail outlets. It is a job which Mr. Paradise admits will take him several years. Although Lord Donald Stokes's decision to retain both Austin and Morris franchises and to give them entirely different model ranges has taken off some of the immediate pressure, there are still too many dealers. Market representation teams are being organized to tackle the whole of the United Kingdom. They will produce a Doomsday Book, similar to Ford's recent operation, listing every detail which can conceivably influence car sales. On their findings Mr. Paradise and Mr Lawrence will make their decisions. Already the distributors have been called in for several long pep talks by Mr Paradise, his staff call it 'indoctrination'.*

Later in the interview there was a positive message on Filmer's market share ambition: 'From a 28 per cent market share in October the division as a whole now holds 31 per cent. Mr. Paradise is adamant that within two years this will be over 35 per cent and this in a market where a maintained one per cent improvement is considered outstanding.'

I don't recall seeing this article or hearing about it, but I agree with one point, disagree with another and partially agree with another. I agree there were too many dealers at that time and some reduction was necessary. However, my disagreement is because growing market share to over 35% was never going to be completed in two years. By October 1971, two and a half years later, new distribution territories would come into being but the dealer rationalization and reduction in dealer numbers would go on well into the 1970s and I and many others would be involved in it. My partial agreement with what Filmer Paradise said is because I appreciate that for the press interview, he could not possibly say the reorganization of the Austin Morris dealer outlets would go on for a very long time.

When I arrived at Longbridge in September 1969 the UK was divided into six sales zones, broadly defined as follows. Zone 1 was Scotland and northernmost England; Zone 2 was northern England and Northern Ireland; Zone 3 was the East Midlands and East Anglia; Zone 4 was the West Midlands and North Wales; Zone 5 was London, the Home Counties and South-East England; and Zone 6 was South-West England, South Wales and the Channel Islands.

It was against this background that I and the other graduate trainees were briefed on the exact nature of the work we were to do. We were to be involved in carrying out a major UK franchising strategy aimed at rationalizing and streamlining the existing Austin, Morris, Riley, Wolseley and MG distribution and dealer networks into one Austin Morris network by the end of September 1971.

Bill Harwood and I would work together on Zone 5 and our first objective was to create and define new geographical distribution territories in Greater London, so that the new territories would be in practice marketing areas based on natural retail spheres of influence. At the time in Greater London the Austin and Riley distributor was Kennings (trading for a while as Kennings Car Mart). The Morris distributors were Henlys, and Stewart and Ardern. The Wolseley distributor was Eustace Watkins and the MG distributor was University Motors.

Each of these distributors had a number of defined geographical territories that collectively gave them exclusive distribution rights in

Greater London. These distribution rights were financially very important and profitable because every vehicle they wholesaled to one of their official retail dealers earned them a 4% commission.

In order to visualize the existing distribution arrangements, we went to 'Arthur the Maps' and he provided us with a Greater London Map together with tracing paper so we could overlay the existing territories on top. Having done so with five overlays for the five franchises it was evident there was no commonality whatsoever and looking down it seemed like we were looking at various overlapping spiders' webs.

The most difficult step was to decide how to create new logical marketing area territories. Not only should they be based on retail centres but we had to calculate the future sales potential so that when presented to the incumbent distributor we could be fairly sure the new territory represented a good and viable business opportunity.

To help us Des Haydon and Richard Wright had recently carried out the first detailed on-territory 'depth study' of Blackburn and Burnley in Lancashire, and as we were able to learn what they had done, we had some understanding of what the outcome for Greater London should look like, even if our initial work was a desk study. The depth study had an important additional element that our desk study didn't. That was to identify all the dealers that had a future with Austin Morris and those dealers which, for reasons of location and/or performance, didn't. There wasn't any way proposals on a dealer's future could be made from the desk.

What we also learned was that each existing distributor and dealer's location had to be plotted as accurately as possible on the map, and represented by a very small sticky-paper dot. The colour of the dot would signify the franchise, so Austin was red, Morris was blue. 'Arthur the Maps' provided all the coloured dots.

With careful research and a degree of trial and error we put together a proposed new territory structure for Greater London. As each territory had to have a major retail centre some marketing data was needed to justify our decisions, and having identified obvious retail centres such as Croydon, Ealing and Ilford, it didn't take us too long to put together an initial set of territory proposals.

By the time the top management of Bernard Bates and Bert Lawrence came to present the proposed future territories to the existing distributors our initial realignment of franchise territories was not greatly changed. And the new territories with associated new legal franchise agreements would come into being from 1st October 1971.

In addition to rationalizing and streamlining the distribution and dealer networks, the Austin Morris franchise strategy was aimed at consolidating the sales volumes and market shares and providing the basis for sales growth.

In the first nine months of 1969, of almost 785,000 new car registrations in the UK, BL had just over 41% market share, while Austin Morris was the market leader with 31%, Ford had just under 27%, and Vauxhall nearly 12.5 %. Rootes Group (owned by Chrysler), maker of Hillmans, Humbers, Singers and Sunbeams, had just over 9 %.

These four manufacturers accounted for almost 90% of total new car registrations in those nine months, with total new car imports barely reaching 10% market share. The top three importers were Fiat with a 2.2% share, VW with a 2% share and Renault with a 1.8% share. No one else had 1% and in that period Honda had just 1065 registrations and Toyota 972. Between them these two Japanese manufacturers achieved just a 0.25% market share.

These figures remind me that when I was writing my thesis on the UK motor industry at university a year earlier in 1968 there was virtually no mention in it of imported cars, and none whatsoever of Japanese motor vehicle manufacturers. The deliberate reduction in the numbers of Austin Morris dealers in the UK from 1970 into the mid-1970s would coincide with the European and Japanese importers' plans to grow their vehicle sales in the UK, which would give them all a requirement to recruit additional dealers.

The dealer rationalization would provide European and more significantly Japanese motor manufacturers not only with terminated Austin Morris dealers anxious to find an alternative car franchise, but also an almost ready-made platform to grow their car sales in the UK, and fairly quickly. Many of the dealers no longer required by Austin Morris were small family businesses which provided convenient local sales and after-sales service and had a loyal customer base, and so were able to convert these customers away from Austin Morris and into their newly acquired importer brand.

Many of these dealers were also keen to show Austin Morris it was wrong to terminate them, but a few years later were actually pleased to be selling vehicles, such as Honda, Toyota and Datsun (now Nissan) which gave them a better business opportunity than the one they had before and one that would also enable them to grow their business.

Early in 1970 I started to think it was time to change my Morris Minor for something a little younger. This thought was also influenced by the fact most weekends I drove to what I regarded as home in Wheathampstead in Hertfordshire. The Morris was only OK at 50 miles an hour so I wanted something faster. After some research in the motoring pages of the Birmingham newspapers I decided I would like to buy an MGB sportscar. Eventually in late spring of 1970 I saw one privately advertised and, accompanied by a few work colleagues to give their opinion, I went to look at it. As we all thought it was in good condition I bought it for £300. It was a blue 1963 MGB soft-top with the interesting registration number 88GOB. Coincident with this purchase my father agreed to sell the Morris Minor. He sold it for £100, the same price that I paid for it.

Around the same time I moved from the terraced house into a large house in Brandwood Road, Kings Heath, which was rented and occupied by a number of Longbridge graduate colleagues. There were six males, including me, who shared rooms and one lady called Ros Bailey, who had her own large room. We all had cars so the drive and the grass verge next to it were overflowing with our vehicles. Before too long we were receiving complaints from the neighbours.

I also had a number of graduate trainee friends who were sharing a rented house in Edgbaston. These included my work colleague Bill Harwood, John King, who I first met the day before starting work, Ian Buchanan and Alan McLeod. A group of us would often meet in a pub for a drink in the evening after work or go out for a meal. It was during one of these evenings that the conversation got onto the subject of football and we all said we played. As a result it was decided we should form a football team and try to get a few games against some local Birmingham teams. The responsibility for organizing a fixture was given to Alan McLeod, with a memorable outcome.

A decision was made to play in an all-red shorts-and-shirts strip and having got the team arranged and kitted out we waited to hear where we would play our first match. Alan said we would be playing on a Sunday morning at Chelmsley Wood, at that time a large housing development, which was built as an overspill town for Birmingham 6 miles to the east.

When we arrived at the football ground, we were shown to a building in which to get changed and we were all a little surprised to see the opposing team were all wearing shirts with numbers on the back. I remember thinking they looked quite professional. The game started and it was clear after just a few minutes that we were playing a team of a much higher standard than us. We did not play badly but even so I think we did well only to lose 5–0. However, that was largely owing to the excellent goalkeeping of Ron Lowe, who had been goalkeeper in the Liverpool University football team.

We found out we had been playing the main Chelmsley Wood football team, who at the time played in one of the Birmingham amateur football leagues. We wondered how this game had happened. Alan then told us he had put an advertisement in the Birmingham press saying, 'BL football team requires fixtures.' I think the Chelmsley Wood team was misled into thinking they were playing the official BL team rather than some BL employees. I see Chelmsley Town as they are now play in Division 1 of the Midland Football League.

Starting in November 1969 Bill Harwood and I got our first taste of work outside the office. We would be carrying out two on-territory studies in Kent, one of the Medway Towns territory, covering Rochester and Chatham, and one of the adjacent Maidstone territory. I still have my copy of the overall study document of the Medway Towns and so I can very accurately describe what Bill and I, two young graduate trainees, produced. I have to say the top management of Austin Morris was fairly courageous in letting us loose in Kent and for having the confidence to let us determine the shape and size of its future distribution and dealer networks, bearing in mind we had only two months' work experience and no field experience when we started out on the study.

I will not go into any detail here on the content of our study or of the study report. However, in the report there is only a performance analysis of Austin, Morris, Ford, Vauxhall and Rootes and only a brief reference to 'imports'. In March 1970 Bill Harwood and I completed the Medway Towns and the Maidstone depth studies and then, having got everything reviewed and approved, it was all typed up, collated and put them together in a small number of hard-covered documents.

Shortly afterwards a meeting was arranged for us to present our studies and the franchising proposals to the Zone 5 management of Ted de Bell and some members of his team. I don't recall any real disagreement with what we presented and proposed, but there was a point in the meeting when in discussing the future of a dealer a Sales Supervisor of the Zone 5 team said that a certain dealer should be given 'the tin tack'.

I remember these words well because I had not heard Cockney rhyming slang before and also because it came from Harry Knopp, an extroverted character to put it mildly, who had not long arrived from Ford. I quickly found out that 'tin tack' meant 'the sack' (or 'fired') in Cockney rhyming slang and some other rhyming slang words would follow. I would work with Harry in Zone 2 in the north of England a few years later, and he provided me with some never-to-be-forgotten experiences.

Looking at the contents of the Medway Towns study over forty years later, and with the benefit of a career in the motor industry, four issues stand out.

First, there was an overambitious plan to grow Austin Morris sales volumes in the study area from 1310 vehicles in 1970–71 to 2065 vehicles in 1974–75, which would increase market share from 26.8% to 28.3% in the same period. This plan was aimed at reversing an actual decline in sales in the period from 1964–65 to 1968–69 from 1728 to 1159 vehicles, and market share from 34.5% to 28.5%. However, there was no reference in the study to any new products or models or any other initiatives that would be necessary to achieve these growth plan objectives.

Second, while there was some reference in the study to the competitors' dealer networks, it is purely in terms of the dealers' location and standard of premises, and there is no mention of dealer competition as a real threat to Austin Morris's growth plans.

Third, there was no comment in the study on the potentially adverse effects of the proposed dealer terminations, and the potential loss of sales if these terminated dealers took on a competitor franchise and converted their customers to the 'foreign' products. It was expected that the remaining dealer networks would be capable of maintaining and growing sales as forecast by the Marketing department, but there was no detail about how all this would be achieved.

Fourth, there was no comment in the study about the continuation of a two-tier distribution policy for Austin and Morris, and the financial impact of Austin and Morris distributors continuing to receive a 4% wholesale commission for every car sold by their respective retail dealers.

In more recent times I am sure there would be a need to include some comment on the strengths, weaknesses, opportunities and threats of the Austin Morris distributor/dealer representation, the competitor dealer networks, future developments in the area and the proposals in the study.

In 1969 Austin Morris sold 288,700 cars in the UK for a 29.9% market share, of a total of 965,400 new car registrations. So assuming an average new car price of £1000 and 100,000 sales by retail dealers the total wholesale commission at 4% per vehicle would be £4 million, or around £60 million in today's money. That is a very expensive way to do business.

It was policy from the very top management that there would be separate Austin and Morris dealer networks and the continuation of a two-tier system, but I think the study could have made mention of the fact that in the future the ideal representation plan would be to have all franchise holders having a direct relationship with the motor manufacturer.

In the 1970s Austin Morris retail dealers would increasingly consider themselves to be second-class citizens, controlled by their distributor, while the new generation of dealers operating importer franchises would be appointed mainly on a direct main-dealer basis.

If we look forward ten years to 1979, the total new car market in the UK was 1,716,275, which represented an increase of almost 78% on the 1969 figure. However, none of this growth was achieved by Austin Morris. In fact its sales were 275,100, giving a market share of 16.0% or 13,600 sales fewer than that it had achieved in 1969. The market leader in 1979 was Ford with 485,600 registrations or a 28.3% market share, compared to 264,000 and a 27.3% market share in 1969. Ford had successfully managed to grow its sales at the same rate as the overall new car market, with the Cortina being its best seller.

In the early part of 1970, a number of executives were recruited from Ford into Austin Morris to provide a more professional and dynamic approach to the sales organization. Trevor Taylor, who became UK Sales Manager, brought with him a number of executives from Ford including Ted de Bell, who would be Regional Manager of Zone 5 (London and the South East), and Geoff Cash, who would have a similar position in Zone 3 (East Midlands and East Anglia). This meant most of the much older Sales Managers, many from the previous British Motor Corporation era, were replaced.

Coincidentally Trevor Taylor set about establishing Regional Offices in each of the six zones in Glasgow, Manchester, Peterborough, Studley in the Midlands, London and Bristol, so that there would be a closer company relationship with the dealers, and better management control and focus on the business locally. It was clear that there could be an opportunity for some of us to work in one of the Regional Offices and this would get us out of the Sales Block at Longbridge. I couldn't imagine working there for more than a couple of years.

After six months as a graduate trainee in early 1970 I received confirmation that I would be a market representation analyst. This was the first step up the organization ladder and there was a small salary increase to go with it.

During August 1970 my MGB was stolen in Marlow, where I had been staying the weekend with Ros Bailey's younger sister Lucilla, having driven her home from the house in Birmingham that Ros and I shared with five others. This was my second car to be stolen. Fortunately, I was able to borrow a company car for a short while and after a few weeks went by I thought I wouldn't see my MGB again.

In September 1970 I had been with Austin Morris for one year and as a result I had become eligible to buy a new Austin Morris through the company on special employee discount terms. I thought that my time owning a sportscar was over and I decided it was time to be more sensible and to economize. The MGB had been fun at times but it was expensive to run. I could just afford to buy a new Mini and so I did. It was a flame-red Mini 1000, with the registration number WOC977J, and the company's discounted price to me was £660.

The Insurance Company was due to pay for my stolen MGB after six weeks had passed, but just a few days before this deadline was reached I received a phone call to say the car had been found in one of the multistorey car parks at Heathrow Airport.

Incredibly my MGB was not damaged and it seems it was taken probably just to go to Heathrow Airport because that was all the miles it had covered in almost six weeks. I was very disappointed to get the car back because I had mentally decided the car had gone for good and in the meantime I had bought a new Mini and was expecting the insurance money on the stolen MGB to help pay for it. Fortunately I was not charged for the six weeks' parking, but I now owned two cars.

In September 1970 Austin Morris recruited another batch of graduates and a few of these came into the Market Representation department. One was Paul Flack, who joined me as a graduate trainee, and together we would be working on Zone 2 in the North of England.

One of the first things I remember about working there was nothing to do with my normal job.

The management at Longbridge had decided there was a need to carry out a physical stock check of all the Austin Morris vehicles held by the distributors in Liverpool, most of which were parked in a very large compound near the Liverpool docks. A number of us were briefed on the project and we travelled to Liverpool.

I remember the trip well for two reasons. The first was because during the day on the docks the weather was very wet and windy. My job, and that of the others, was literally to open and lift the bonnet of each of the hundreds of cars parked there and check its vehicle chassis number. Having done this I had to find it by thumbing through page after page of a big computer print-out and having found the long chassis number I had to tick it as 'found'. Doing this in the wind and rain wasn't pleasant or easy. The task lasted several days because of the large number of cars we had to check. There was a second reason for remembering the trip. When we had finished the physical checking we had to write up our findings. We were provided with a sales office at

Voss Motors, the Austin distributor in Hanover Street in central Liverpool. We were working past five o'clock and when we decided it was time to leave we realized we were alone.

All doors to the rear of the offices were locked so we walked into the showroom but the doors were also locked. We were locked in the building. It felt like we were trespassing and shouldn't be in the showroom. I can't remember how we were rescued but that evening we managed to phone someone and after a while a person came to let us out.

At Longbridge in September 1970 I was using a pool car and was told I would soon get this as a permanent company job car. At the same time I still owned my MGB and a new Mini. To potentially run three cars was financial madness. Ron Lowe expressed interest in buying my Mini, so I sold it to him for £600. A few weeks later a white Riley Kestrel 1300 automatic four-door, registration UOF187H, became my first official job car. Now I had a company car I no longer needed the MGB. At the end of 1970 my father sold it for £300, the same price I had bought it for.

At some point in 1971 we had to move out of our rented house in Brandwood Road, Kings Heath. The complaints from the neighbours intensified regarding the large numbers of our cars that spilled over from the driveway onto the grass verge adjacent to the main road. Some of us moved to a rented house in Anderton Park Road, in the Moseley area of Birmingham. My main memory is that it was a cold and damp place.

I and a number of my flatmates and work colleagues would meet up for a drink or two a few evenings a week in the White Swan in Harborne in the suburbs of Birmingham. It was a popular place to meet and always full of young people. In late October 1971 and towards the end of one evening in the White Swan I was approached by a very attractive young lady, who put a very small piece of paper in my shirt pocket. No words were said.

The piece of paper just had a phone number on it. I had a phone number but no name. I was very interested in this young lady as we had exchanged looks in the pub and she obviously had some interest in me. I was intrigued and decided to look through the Birmingham telephone directory to see if I could find a surname and an address. I literally went through the whole directory, but could not find a name to match the phone number.

I wanted to speak to the young lady so I had no option but to phone the number. When I did I spoke briefly to a man with a strong Eastern European accent. I found out later he was the young lady's Polish father. He wanted to know who I was. I left a message to say I hope to meet your daughter again, and from this I did meet this young lady a few days later in the White Swan. It was no wonder I couldn't find the phone number in the phone book because I found out later it was ex-directory.

A few weeks later I was very pleased to receive the offer of a promotion based at the Regional Office in Sale, Manchester from January 1972. This would enable me for the first time to have a fully operational role away from the Sales Offices in Longbridge.

By the end of 1971 the young lady, called Annette, and I had become very close and so we had to decide how to handle my job move north. It was clear Annette wanted to go to Manchester with me and I wanted her to go with me. We knew her Polish father would not allow her to go with me unless we got married, so my move to Manchester forced the issue about our relationship. We decided to get married. We got engaged to be married on 22nd January 1972, and planned our marriage for 16th September 1972.

Chapter 4

# British Leyland 1972–1974

**THE REGIONAL OFFICE IN SALE** was situated on Cross Street, the main A56 road on the southern outskirts of Manchester. Immediately behind the office and attached to it was an engineering works and for much of the day the noise of metal being cut greatly interrupted what could have been a fairly quiet work place.

During the week I stayed at the Normanhurst Hotel in Sale. This was a small family-run hotel within our company's expense policy limits for overnight accommodation and evening meals. Austin Morris Service and Parts representatives often stayed there and we usually had our evening meal with the hotel owner and his children.

You don't tend to remember anything years later about staying at a hotel unless it is very unusual. One evening at the Normanhurst I met an Invicta Airlines crew drinking at the hotel bar. I asked them where they had been, assuming they had arrived at Manchester Airport and were staying at the hotel for the night. The reply surprised me. 'No, we are going to fly soon.'

At the same time I started working in Sale, Annette and I started looking at houses to buy within a reasonable commuting distance of the office. There were limits on what I could afford to purchase because the maximum mortgage I could get was two and a half times my annual salary. After some searching we found and bought a three-bed detached bungalow being built in Sandbach, Cheshire. The price was £6695 and I put a deposit down on 14th February 1972. It was a convenient half an hour up the M6 and A56 to the Sale office. I got the keys to 7 Gawsworth Drive on 24th May 1972 and moved into my new home.

One of the main business events in the regional office was the monthly wholesale meeting, which the field force of sales supervisors and office-based sales staff attended. All through the early 1970s there was considerable pressure on the sales supervisors to wholesale and allocate the total volume of vehicles scheduled to be produced a few months ahead. The simple objective was that every car and light commercial vehicle to be produced would have a distributor name and location on it when it was ready to be dispatched.

However, back in 1972–1974 the exact number of each vehicle and model type to be produced was mainly determined by management at the factory, so it was a case of distributors and their dealers selling what the factory was going to make, rather than the factory making what the distributors and their dealers forecast or expected to sell. About the only choice the dealers usually had was which body colour to order.

So each month each sales supervisor would be required to get each of their distributors to agree to order a predetermined number of vehicles which included Minis, Austin 1100/1300s, Austin Maxis, 1800s and MGs, which by now all Austin Morris distributors were able to sell. The 1100/1300 was replaced by the Allegro, the Morris Marina was added and the 1800/2200 range was replaced by the Princess. The wholesale process also included Mini vans and Morris Marina vans. There were also JU and J4 commercial vans to allocate to some distributors until replaced by the Sherpa van in 1974. The Austin Morris model range also included Austin Taxis which were wholesaled to Lookers for its exclusive Austin Taxi dealer Cross Street Garage in Manchester.

When road-tested by *Autocar* magazine, a 250 JU's maximum speed was 58 mph. Testers found it 'clumsy to drive' with the 'need for constant expertise to overcome its faults'. Engine noise, especially above 40 mph (64 kmh), was a source of criticism. The Morris J4 also gave a poor driving experience even by the standards of the day. Neither the JU nor the J4 sold well and could not compete with the Ford Transit.

All the numbers of vehicles in stock, on order and required to be ordered were detailed on a large wholesale form of A3 size, and once agreed and signed by the distributor principal or manager, he committed his company to order exactly the number of vehicles as detailed on the form, by a fixed date also shown on the form. It was no wonder with so many models and model derivatives that the wholesale form had to be of A3 size.

The predetermined 'allocation' of vehicles, model by model and for all main model derivatives, assumed that all distributors and their dealers achieved at least 100% of their agreed monthly sales objectives. As 100% of sales objectives were not always achieved this often meant the sales supervisor was required to obtain a level of orders greatly above the numbers his distributors wanted or needed to order. This was because the distributor usually had enough vehicles in stock plus vehicles already on order to meet its projected sales levels, without needing to order any more vehicles.

Requiring distributors to order more vehicles than they wanted or needed meant the monthly wholesale process became a game of hard negotiation between the sales supervisor/manufacturer and the distributors. The more desirable cars such as the MGB sportscar were often offered to the distributor on condition the less desirable models were also taken and ordered, so achieving a compromise wholesale 'package' acceptable to both parties.

On most occasions at the end of the wholesale meeting, the final message from management to the sales supervisors would be 'Don't come back until you have got 100% order take.' No pressure there then! A few weeks later the whole sales team would gather again in the Regional Office for what was called a 'tie up' meeting, where hopefully the final total order rendition would be discussed and signed off.

The meeting would also review the zone's sales performance because every ten days the sales figures would be available model by model in some detail. In the early 1970s Austin Morris's number-one position over Ford was under serious threat, with its lead in market share and actual sales over Ford fast reducing.

However, despite the continuous sales pressures I thought there was a very good team spirit in the Zone 2 office. After the morning wholesale meeting ended, we would usually walk a hundred yards down the road to an Indian restaurant for an informal and often light-hearted lunch.

At the end of lunch, it fell to the youngest among us, who was the sales analyst, to collect the right money from everyone, pay the waiter and then get enough lunch receipts for everyone who could claim expenses for lunch. The lunch allowance in 1972 was 65p.

The dealers were graded 'a', 'b' and 'c'. Dealers graded 'a' were considered to have a long-term future, those graded 'b' had a future subject to development and dealers graded 'c' were those without a future and scheduled to have their franchise agreement terminated. The 'c' dealers were usually very small family-run garages selling low numbers of new cars. There was an Austin or Morris dealer in almost every location where there was some population, and there were also some just selling a few MGs a year.

There was a form of words I had to use on these 'c' dealer visits. In summary I had to tell the dealer principal that a detailed franchise review had taken place to determine the future dealer network requirements in the area and unfortunately Austin Morris had decided it would not be able to include the dealer in its future plans. The dealer would be given the appropriate period of notice of the termination of the franchise agreement, normally one year, and this would be confirmed in a letter. On these dealer termination visits I was accompanied by the distributor under whose control the dealer operated.

I remember my first dealer termination meeting in February 1972 largely because of the response I received after I had delivered the bad news. The dealer was Kings of Oxford, which was located literally next door to its Austin distributor, Lookers of Manchester on Chester Road, Old Trafford. In view of its location the bad news I had to give should not have come as a big surprise. However, while I could understand the dealer principal would not be pleased with what I had to say, the response was more aggressive than I expected. It concluded with some table thumping and with me being asked the name of my boss, then who was his boss, then who was his boss, until we reached the ultimate BL boss Lord Stokes. I suppose you could call this experience a 'baptism of fire'.

This first dealer termination visit was very helpful because I realized I had to expect almost any type of response and reaction from a dealer being told it had no future. I would come to experience all sorts of responses and reactions.

In March 1972 I had to replace my company car, an Austin 1300. I had enjoyed driving this car since June 1971 especially as it had proved to be comfortable and reliable on my travels. Now, however, I took delivery of a Morris Marina 1.3 four-door saloon in Russet Brown, a chocolate-brown body colour, with brown vinyl seats, registration COB592K – which proved to be less reliable. Just a few weeks later the fanbelt broke and my wife-to-be very kindly gave me her tights to use as a makeshift one. Many Marina owners used this emergency method when fanbelts broke.

The Marina's metal body was badly protected and poor body and window seals allowed water into the interior and boot. Quality was not designed or built into the Marina and one service fix was for dealers to drill a few holes at some specific points in the boot. Apparently the rationale was that by doing this more water would get out than get in! The Marina has been described by many as one of the worst cars of all time, but it should be remembered that it was one of the best-selling cars in the UK throughout its life from 1971 to 1980.

Travelling to and from dealers meant I spent many hours on my own in the car. The company cars we drove in the early 1970s only had a radio, but this at least provided some entertainment to alleviate the boredom as I could listen to almost continuous pop music on BBC Radio 1 or daily news on Radio 4.

Almost all of the visits to issue notice of termination were to small family-owned retail dealers. One visit however was to an Austin distributor in Blackburn, Lancashire called Townley Motors Ltd. My meeting with Mr Barton Townley, the owner, was prompted by some sale or proposed closure of the premises, which meant the franchise could not continue.

Mr Townley told me he thought I was rather young to be doing what I was doing, but the meeting was memorable for me because this elderly gentleman had seen the whole evolution of the UK-based motor industry through the twentieth century. Indeed, he told me he had played an important role some forty-plus years before my visit, by helping pay the wages of workers at the Austin Motor Company during a time when the company was in financial difficulty.

Many of the dealers I visited to confirm the end of their franchise agreement with Austin Morris were important first-generation motor traders who collectively had helped make Austin Morris the number-one motor manufacturer in the UK in terms of vehicle sales. I was very aware that I was helping to bring one generation of Austin Morris dealers to a close. However, I was equally aware that many of these dealers would go on to take competitor franchises and by doing so hasten the decline of Austin Morris car sales in the UK.

As I became more experienced at delivering 'termination notice' to some dealers I tried to adopt a professional yet sympathetic style of delivering the 'bad news'. On one occasion at the end of a short meeting one dealer principal called me a nice assassin. I think it was a compliment.

I was never totally sure how a meeting with a dealer would go. One afternoon I visited a dealer to find that the dealer principal was a lady and she ushered me to a very small office in the corner of the car showroom. As usual I explained details of the franchise review and that this dealer's franchise agreement would be terminated. I was not prepared for what followed. The lady started to cry. Once she had regained a degree of composure she told me her father who had been the owner of the company had died a few weeks earlier and she was hoping to continue the business. I found this situation difficult because even though we had not been informed of the owner's death, the lady clearly did not know the reason for my visit and what I had to say was 'a great bolt out of the blue'.

Occasionally we were looking to appoint a new dealer. It was quite normal for the dealer candidate I was pursuing to say they wanted the Rover Triumph franchise in addition to Austin Morris, especially if I was trying to get the dealer to give up their existing franchise. These requests for Rover Triumph brought me into contact in 1972 with John Cunnane, who at the time was the Regional Manager responsible for the Rover Triumph division in the north of England. John was rarely able to help with his franchise, but he would become an important influence on my career in Land Rover Ltd in the 1980s.

Ron Parr was the Rover Triumph Regional Sales Representative reporting to John Cunnane and I can recall several enjoyable dealer visits with Ron when he would also recount amusing stories of his time 'on the road'. Ron told me one story that he claimed was true and I repeat it here. On one of his dealer visits Ron arrived at the dealer's showroom to be told the meeting would be in the dealer owner's house next door and that he should go there. On knocking on the front door, a lady came to the door without any clothes on and invited him in. Ron did not tell me what happened next, but with a smile and a laugh he told me it was not Rover Triumph business. Ron would later work for me in Land Rover on Fleet Sales and I wish he were alive today to verify this story!

Although both Austin Morris and Rover Triumph, which included Land Rover, were part of BL there were in practice very few opportunities to produce a combined Austin Morris and Rover Triumph franchise 'package' for any dealer 'conquest' proposal I was working on, and which would also benefit Rover Triumph.

Similarly, I would often be asked by a dealer candidate if I could explore the possibility of them taking on the Jaguar Daimler franchise. This meant contacting Bill Benton, who was responsible for the northern half of the UK. I can't recall any situations where Jaguar Daimler could help me and I think the attitude within Jaguar was one of not wanting to be operationally associated with Austin Morris, as we were seen as 'metal movers'.

The three divisions of Austin Morris, Rover Triumph, and Jaguar Daimler were working independently in the early 1970s and I did not see any evidence of interdivision cooperation at the regional level. All this would change in the mid-to-late 1970s as the result of the 1975 Ryder Report, more of which later.

So 1972 was a busy year, because in addition to having a new job in the north of England and buying my first property Annette and I had fixed our wedding for 4.30 p.m. on Saturday, 16th September at St Augustine's Church, Edgbaston, in Birmingham. The evening before the wedding I stayed with Annette's Aunt Janet and her Uncle Johnnie at their house in Sheldon on the eastern side of Birmingham. Johnnie insisted on taking me to a nearby pub that evening and also at lunchtime the following day. In the early afternoon I told him I wasn't going to spend more time in the pub and I went for a walk on the path alongside the A45 Coventry-to-Birmingham Road. Here I encountered a number of friends and relatives on the way to Edgbaston and I gave some of them a few words on directions.

The weather that day was kind to us for the photographs taken that afternoon, and after the wedding a reception was held at a hotel on the Hagley Road, Edgbaston. Later that evening Annette and I left and with gravel clanking inside the hub caps of my Morris Marina as we set off to Coventry, where we spent the night at the Post House.

We then flew from Luton for two weeks in Athens. Our hotel was in Glyfada on the coast south of Athens. While there we took a full day's boat trip to the Greek islands of Hydra and Aegina. We also visited Cape Sounion, south of Athens which is noted as the site of ruins of an ancient Greek temple of Poseidon, the god of the sea in classical mythology. The ruins bear the deeply engraved name of English poet Lord Byron. While in Athens we went to one of the best beaches in the area at Vouliagmeni and unbelievably one day we met my best man Jim Obee there. We didn't know he was in Athens and he claimed not to know we were there. Some coincidence!

On Saturday, 2nd February 1974 at 5.30 a.m. my daughter Gillian was born at Leighton Hospital, Crewe. I had been with Annette on the Friday evening and in the early hours of Saturday I was told to go home and get some sleep. I was awoken by a phone call barely a few hours later to say I was the father of a baby girl. Annette was in hospital for another four days until I brought her and Gillian home on the morning of 7th February.

One of the highlights of the working year for some younger members of the sales organization was the annual Motor Show, which in the early 1970s was held at the Earls Court Exhibition Centre in West London. As the leading motor manufacturer Austin Morris had a very large stand in order to display almost every model produced and those of us working in the Regional Offices were required to man the stand. Each year we would be measured for a new suit, and provided with a few shirts, a tie and a badge, so it would be clear to visitors we were Austin Morris representatives.

On arriving at the Austin Morris stand, pads of sales enquiry forms were issued, and for each enquiry taken we had to write down the customer's name, address, phone number, existing car and the model they were interested in. These three-part forms would then be separated, with one part being sent to the appropriate dealer for follow-up with the prospect. Each morning before the Show opened there would be a briefing from management about the level of enquiries we had taken the previous day with some words of encouragement for the coming one.

There was one ploy adopted by some on the stand to improve the number of enquiries they were taking. This was to say to a visitor wanting a brochure or price list that we had temporarily run out, but if you give me your name and address we will send you one. These were to be counted as a real enquiry but dealers following up the enquiry later invariably thought the enquiries were not genuine and often a waste of time. However this ploy ensured those who used it could achieve a reasonable number of enquiries without too much effort.

I first attended the Motor Show at the Earls Court Exhibition Centre in 1971 and I spent a week on the MG stand looking after the MGB, MGB GT, MG Midget and MG 1300. It was a good stand to be on because the MG cars attracted visitors who were genuinely interested in sportscars and sporty saloons and which were more glamorous than Austin and Morris cars. Most of us stayed at the Cumberland Hotel near Marble Arch, which was the first time I had stayed at a big London hotel. This was an enjoyable week because we all went out in the evening for dinner and some drinking, and we were away from normal office work.

For the October 1973 Motor Show, I was responsible for the Austin Allegro. The Allegro replaced the popular Austin 1100 and 1300 and was launched a short while earlier in the year, so this was its first showing at the Motor Show. I don't recall visitors being excited or gushing with praise about the new Allegro.

'It's a car which we think will appeal not only to the sophisticated British public but to the sophisticated European public', said BL's MD Lord Stokes rather optimistically on its launch. Unfortunately, while it was not a bad-looking car, the Allegro became infamous for its square steering wheel with rounded corners, called 'quartic'. This shape was meant to create some extra space between the bottom of the steering wheel and the driver's legs. However, the 'quartic' wheel was not popular and it was replaced with a conventional round steering wheel two years later.

In August 1973 my new company car was an Allegro two-door in gold body colour with the registration GOK225L. The cars for the sales supervisors, including mine, were preproduction vehicles, and everyone in the field force had one with a similar GOKxxxL registration number, so they became known by those that had them as 'Gockles'. I don't think many enjoyed driving the Allegro – I certainly didn't. The Austin Allegro was meant to be a saviour for the company, but it was soon apparent that Austin Morris would find it difficult to hold off the challenge of Ford with its Cortina and Escort models, as this depended largely on the performance of the Allegro and Marina, which were less favoured by companies and fleets.

The Manchester metropolitan area had the benefit of having the Altrincham Regional Office within its boundary, so I could do all my administrative work there. I was responsible for achieving the sales objectives of six Austin and six Morris distributors in a geographical area covering Bolton, Bury, Oldham and Rochdale to the north, the main Manchester and Salford conurbation in the centre and Stockport, Wilmslow and Macclesfield to the south.

The Austin distributors were Lookers, located at Old Trafford, Macclesfield, Rochdale and Stockport as well as Henlys in Manchester and Southern Brothers in Bolton. The Morris distributors were Lex Motor Co. located in Bury, Manchester, Rochdale, Stockport and Wilmslow, plus Kennings in Manchester. I also had one major commercial vehicle distributor, namely Kennings Truck Centre in Manchester, which retailed light commercial vehicles as well as Leyland Trucks, and one Austin Taxi dealer. Between them these thirteen distributors and forty-seven retail dealers totalled sixty outlets in my area of responsibility.

I had to visit the distributors at least once a month to carry out the monthly wholesale procedure face to face with the appropriate management. The initial task I had each month was to break down the total allocation of vehicles I had been given, which was well over 1,000, so that I could offer to each distributor a fair and equitable number of vehicles by model. I would spend several hours on the evening of the wholesale meeting at home calculating the numbers for each distributor and then entering them on the large A3 wholesale forms. The whole process was very laborious and time-consuming, but that was how we did it at that time.

I would start my round of distributor meetings on dates and times I had already confirmed. There were usually three or four meetings a day, and often a second meeting before that month's wholesale was agreed and signed off. One operational issue for me was that Lookers took a large share of the total Austin 'allocation' of vehicles and Lex, representing five of my six Morris distributors, took an even larger share of the Morris 'allocation'. This meant I had very little flexibility, so if Lex as a group decided it did not want to order the total allocation of the Morris Marinas on offer, I only had Kennings available to help with orders.

I found it was important to develop a good working relationship with the distributor management and that any negotiation needed a degree of cooperation and compromise. That form of negotiation worked well for me and generally for my distributor colleagues too. Adopting a hard dictatorial approach and insisting on the number of orders rarely worked for very long and was not conducive to a long stay in the role.

While the 'packaging' of desirable models with the less desirable could work on occasions, I learned that one or two sales supervisors often took adventurous and slightly desperate routes to get their wholesale ordering numbers agreed with their distributors. On one occasion a sales supervisor played a game of darts with the distributor's wholesale manager over a disputed number of cars to be ordered, and another did the same, with the winner of a game of snooker deciding on some vehicle orders. I was told these stories were true but I never thought this was good business practice and tried to be as professional as possible in carrying out the wholesaling exercise with my distributors.

I often had to 'sell' showroom merchandise. One example was a life-size freestanding cardboard cut-out figure of Sir Alec Issigonis, who had designed the original Mini. As a cardboard figure he had his hand stretched out so it could be placed on the roof of a Mini in the showroom.

One dealer I remember well was Hymie Wernick, the owner of Collins Autocar in Prestwich, in north Manchester. When I went there, he had new products, such as electrical goods, which he tried to sell me at a special price, but I was able to resist his offers. I learned a lot from him about the used car business, and the legal and not-so-legal activities that went with the buying and selling of used cars.

The Sales Supervisor's job was a great way for young field personnel such as me to learn about the motor vehicle industry at the 'sharp end' of the business. It was also valuable for me to have had this field role experience, as later in my career when I had responsibility for a sales field force I knew what the Regional Managers were meant to be doing and could make constructive comment, having done the job myself.

I noticed the big operational differences between the BL brands one day when I was visiting Henlys, one of my distributors in Peter Street, Manchester. I happened to bump into Bill Benton, the Jaguar Sales Representative, who was visiting David Rosenfield, the Jaguar Daimler distributor, which was also situated in Peter Street. I had a brief conversation with Bill when he told me he was going to lunch with his distributor at the luxury French restaurant of the Midland Hotel, also conveniently located in Peter Street. I replied by saying I would be having lunch with the Henlys management in the Midland Hotel Coffee Shop!

In the early 1970s Austin Morris's market share was gradually falling, not only under strong competition from Ford, but also the increasing competition from imported car brands from Europe and Japan. The large number of dealers terminated by Austin Morris fuelled the growth of Japanese dealer networks, especially those of Datsun (later renamed Nissan), Honda and Toyota.

In 1974 top management decided a major sales initiative was necessary to stop Ford taking its number-one sales position. This initiative took the form of what I believe was the first big company sales campaign in Austin Morris where large financial incentives were available to distributors and dealers for achieving specific sales volume targets.

The campaign was broadly as follows. Distributors and dealers were paid financial bonuses for achieving vehicle registrations up to and above a total volume target in a fixed two-month period. There was also an incentive for Austin Morris's sales supervisors achieving their vehicle registration target. This was an Austin Reed suit voucher. There was a great deal of pressure put on the field force to achieve at least 100% of the target and the top management's message quite simply was for us to get the maximum number of vehicle sales and registrations.

I recall a meeting on this subject with Ken Martindale of Lookers Manchester, one of my Austin distributors. Our discussion led to a large number of Manchester Police Minis being registered earlier than planned to take advantage of the campaign bonuses. This greatly helped Lookers and my area sales results in the campaign.

My wholesaling activity each month of hundreds of Austin vehicles, including the Allegro and Maxi, was challenging to say the least, so Ken's support contributed significantly to my wholesaling results and was essential to my area's Austin sales performance. As it was my first field sales job, I also learned a lot from Ken about the car retail business. My distributors did a great job in bringing forward sales and registrations into the campaign period, and my area's results were in the top few nationally. I received my Austin Reed voucher and bought one of these fairly expensive new suits soon after.

There was however a sting in the tail of this first major sales campaign. With the help of my distributors, I had pulled forward every piece of business I could, especially fleet sales, in order to obtain the maximum number of registrations. The campaign was considered a big success and I remember being congratulated by the sales management. However, as soon as the campaign was over and we were into the next month it was no surprise that my area's vehicle sales and registrations were abnormally low. And I was not alone. I was asked to explain why the figures were so poor. I didn't need to explain the reason and I thought it was unfair to criticize the field force for a poor month's sales. It could have been easily predicted!

There were several lessons I learned from this campaign that I would remember at times later in my career when considering sales action plans. First, you can only sell (register) a new car once, so it was inevitable that achieving an artificially high number of registrations in one month by offering high financial bonuses would lead to a sizeable equivalent reduction in registrations the next month. Second, it was financially an extremely costly way to sell cars and do business. Third, a campaign structured as this one was and which would be costly for the manufacturer, although not planned, could be financially very beneficial for a distributor and dealer network. Fourth, a strategic objective to outsell and beat your competitors could not be successful if it was delivered with a very short-term tactical initiative.

Not long after this sales campaign ended my time as a sales supervisor also came to an end.

# Chapter 5
# British Leyland 1975–1978

**AT THE START OF 1975** I was informed that from 1st February I was being promoted to the position of Franchise Development Manager for the Austin Morris Division in Zone 3, which covered the East Midlands and East Anglia. Initially I wasn't keen to go to Peterborough as Annette and I had settled in Sandbach, we had daughter Gilly not quite one year old and I was enjoying the sales supervisor role.

However, it was my first promotion into management and in those days if you wanted to progress your career a promotion like this couldn't be refused. You went where top management wanted you to go and this normally meant a relocation of you and your family to another area. I realized I could not refuse the promotion and so I went to Zone 3 in Peterborough.

As Franchise Development Manager I reported to two managers, functionally to Bob Blackshaw, Franchise Operations Manager, based at Longbridge, and operationally to Chris Martin, the Regional General Manager, based at Peterborough. This double reporting was not unusual at that time and it was no problem to me. I would also have two persons reporting to me, Tony Darlison, a Business Management Consultant, and a Franchising Coordinator, to be appointed.

A major benefit of the move was a salary increase of around 20% for entering management starting at grade 25 and the company would pay for all expenses incurred in the move, especially those costs of selling my existing home and buying a new one. My management grade was 26 and my appointment letter stated that 'in line with company policy payment for overtime working does not apply at corporate grade 25 and above'. The company knew how to get value from its management!

The Zone 3 Regional Office was on Lincoln Road, just north of the centre of Peterborough, on the main road running north to south through the town. The office itself was not big and was situated above a moderately sized supermarket, and we shared the car park at the back. The office was largely occupied by the Austin Morris team but also included a small number of Rover Triumph staff headed by John Evans, the Regional Manager.

The move to Peterborough meant I started each week with a long journey from home in Sandbach, down the M6 motorway to Birmingham and to Lutterworth, where I continued on slower roads through Market Harborough, Corby and Wansford. I still had an Austin Allegro which wasn't the best car for such long journeys. Three hours later at around 9.15 a.m. and after 145 miles I arrived at the Peterborough office.

In my first week in Peterborough I was booked into the three-star Great Northern Hotel which was situated immediately opposite the Peterborough railway station to which its ownership was linked during its early life. According to records the hotel opened on 1st April 1852. When I stayed there it was old-fashioned, but it was one of the few decent hotels in the town that I could stay in within the company expense policy limits.

The hotel had a standard non-changing dinner menu that was also limited in choice, but acceptable, considering there were no good restaurants as an alternative that were worth going out to. So at the end of the day I went to the hotel, got changed, had dinner and often worked, as the single rooms I stayed in did not have a television.

The Ryder Report became public on 23rd April 1975. What was it and why was it significant? Here is a brief summary. By the end of 1974 after many difficult years British Leyland (BLMC) was near to bankruptcy. Its financial backers persuaded Lord Stokes to approach the Labour government for financial assistance. The end result was that the government agreed to guarantee BLMC's growing overdraft with the banks in exchange for a role in running the business.

The National Enterprise Board, led by its Chairman, Sir Don Ryder, was appointed on 18th December 1974 'to conduct, in consultation with the corporation and the trade unions, an overall assessment of BLMC's present situation and future prospects, covering corporate strategy, investment, markets, organization, employment, productivity, management/labour relations, profitability and finance; and to report to the Government'.

The Ryder Report when it was published on 23rd April 1975 recommended that the company maintain a presence in both volume and specialist manufacturing and that it should be reorganized into four divisions, namely BL Cars, BL Trucks and Buses, BL International and BL Special Products. Although Ryder recognized that the car division's brands should retain their independent identities, he suggested the company had to pool and integrate its resources.

The government gave the plan its full and unconditional approval and as a result on 27th June 1975 BLMC ceased to exist as an independent company and became known as BL Ltd fully under government control. The National Enterprise Board was given the task of overseeing the direction that BL was going in, and ensured that Ryder's plans for the company were being implemented. This government control would have significant implications for the company over the coming years as successive governments appointed and then replaced the very top management as the financial problems were not resolved.

The Ryder Report, as with previous company plans, was very optimistic about the future of the company, especially in its forecast that BL would maintain a 33% market share in the UK. However, as I noted at the time, the Report made no detailed proposals to improve the manufacturing productivity or product quality that would be necessary to improve the competitiveness of the business. The changes to be implemented were primarily to the organization and management structure.

Ryder recommended the expansion of BL as a means to achieving financial stability and a secure future but the unpopularity of the company's products coupled with increasing competition, trade union unrest and the resulting lost car production would lead to failure rather than success of the Ryder Plan.

The main implication for the existing sales and marketing organizations and for managers like me was that we would quickly see the start of a progressive integration into one structure of the BL brands that had hitherto been run separately and independently. For me the Ryder Report would shortly mean a significant change in my job profile, and with it increased operational responsibilities.

In May 1975 I finally handed in my Austin Allegro, which I was very pleased to do, and collected a white Austin 1800 four-door saloon, registration number HON308N. The body colour was white, but it was a preproduction model and the body panels were of various shades, which were clear to see. It was not a great advertisement for the company or for me with a role aimed at improving the quality of the dealer network.

The car was launched on 26th March 1975 as the 18-22 Series: 'The car that has got it all together.' It was a bold claim, one soon to be clearly overpromised and underdelivered! For the first six months of production three badge-engineered versions were produced for Austin, Morris and Wolseley. By September 1975, all the models in the range were sold under the Princess name and the Wolseley marque was dropped.

On the 6th June I made an entry in my diary that I had a meeting at Longbridge with Bob Blackshaw regarding my job, and it was the first indication I had it would expand to cover all BL's car brands.

Later in June I interviewed someone who was recommended to me by management at Longbridge for the vacant Franchising Coordinator position. I was not impressed and tried to stop the appointment taking place but I was told I would have to take the person. I had more important matters to handle and so I gave way.

In July 1975 communications within the company centred on the reorganization of Austin Morris, Rover Triumph and Jaguar Daimler, with these divisions now being integrated into one BL Cars structure. What happened in practice was that Austin Morris Regional Office management became BL Cars management almost overnight. Those who were in the Rover Triumph and Jaguar Daimler Divisions saw the reorganization as quite simply an Austin Morris takeover.

In the Zone 3 Regional Office in Peterborough, Chris Martin became General Sales Manager for BL Cars with responsibility for all the company's vehicle brands, namely Austin, Morris, Rover Triumph, Land Rover, Jaguar Daimler and Sherpa vans. Roger Charlton moved from Bristol to become Field Sales Manager BL Cars, and I was now Franchise Development Manager BL Cars. John Evans, who had been the Rover Triumph Regional Manager, took on the role of Light Commercial Vans Manager, responsible for Sherpa van sales. The Regional Service and Regional Parts Managers in Zone 3 were David Greaves and David Lansdowne respectively. Bryan Tunstall joined me from Rover Triumph as a second Business Management Consultant.

One other change was that I moved into what had been the Rover Triumph office and acquired a secretary called Joan Lee. However, my new office was quite small and Joan's desk immediately adjoined and faced mine so if I wanted to make a phone call she had to stop work on her typewriter. Around mid-morning and mid-afternoon Joan would get the china cups and saucers out of a cupboard and I would get tea and biscuits, following in the Rover Triumph tradition. At the old Austin Morris end of the office, the use of plastic cups for tea and coffee continued.

While all this reorganization was taking place it had been difficult to sell our bungalow in Sandbach and I was therefore continuing long journeys to and from home to Peterborough with two or three nights a week in the Great Northern Hotel. However, from July Roger Charlton was in the process of relocating from Bristol and like me was spending a few nights in the Great Northern. This made a welcome change from having dinner alone.

That month I made my first trip to Lincoln to meet one of the long-standing Jaguar Daimler distributors, owned and run by Reg Mansbridge. He was greatly concerned at the possibility of losing his Jaguar franchise to either Wrights or Gilberts, the two Austin Morris distributors in the city. Reg told me he had acquired the Jaguar franchise in the early 1950s largely as the result of competing with his wife Joan in many of the major motor car rallies of the time, such as the 1953 Rally Coupe des Alpes, in his Jaguar XK120 FHC. The refranchising activity I was soon to be involved in would not be like that this time.

It would be some time until the new Leyland Cars franchise policy would be confirmed and so these initial meetings were mainly to learn and understand the nature of the Rover Triumph and Jaguar Daimler specialist car networks. Meetings with Reg Mansbridge were always cordial and beneficial to me as he spoke frankly about a whole range of Jaguar-related business matters and I remember our meetings included a fine lunch at the White Hart Hotel in Lincoln.

On 15th September Annette and I finally sold our bungalow in Sandbach and could now progress the purchase of our new house in Kesteven Close, Market Deeping, a small town 8 miles north of Peterborough. We signed contracts on 24th September, but did not complete and get the keys until 13th October. It brought to an end over seven months of a long relocation process which meant too many nights in hotels away from my wife and very young daughter and too many long return journeys from home to work.

Coincidentally, Roger Charlton moved into Elm Close, the next cul-de-sac to ours, and David Greaves was already living just a few houses away in Elm Close too. The Austin Morris dealer in Market Deeping was Todd's Garage and I often wondered what Mr Todd thought about the presence of three of the motor manufacturer's managers, responsible for sales, service and franchising, living so close to his business.

My new Franchising Coordinator had been with me one month when it was clear to me he was unsuited to the position. I am not naming him directly because for me that doesn't matter compared to my unforgettable experience with him, but I will call him Mr X. Until now Mr X had just been working in the office on franchising topics but he didn't grasp the basics of the job and I couldn't see him implementing or being trusted to implement on-territory franchising actions with dealers on his own.

It was decided Mr X and I would visit some dealers so we could make an informed judgement about his suitability to continue in the job. I had arranged for the two of us to visit some dealers in Essex over two days, which meant an overnight stay there and that we should meet at the Harpenden Moat House Hotel in Hertfordshire. We agreed this was a convenient meeting place as he was coming from the Midlands down the M1 motorway.

The plan was that having met up in Harpenden I would go with Mr X in his company car and my wife Annette would take mine. We were staying with my parents just before moving into our new house. The meeting time duly came and went and it was over an hour later before Mr X showed up. Not having mobile phones in those days made communication difficult in situations like this. I asked him where had he been and he said London. That was not all because he said he had to pawn the car's tax disc to get some petrol. That I found hard to believe, but he said he had not brought any money with him so he had no choice.

I asked him how he was going to pay for his hotel bill and other expenses incurred during a two-day field trip. No sensible answer came back. During the rest of the day we travelled to Rayleigh, Benfleet and Canvey Island to do a short franchising study. This was fairly uneventful except when we were going downhill even on main roads as Mr X decided to switch off the car's engine and coast down the road. He said it was to save fuel.

The following day we had distributor meetings in Southend. All I remember on that day was being a passenger in a car being driven by Mr X on the wrong side of a dual carriageway towards oncoming traffic on the A127 Southend Arterial Road. Although it was only for a short distance it was very unnerving to say the least. This trip convinced me Mr X was unsuited to carrying out the on-territory aspects of the job and I insisted on a meeting to bring this unsatisfactory appointment to an end as soon as possible.

A few weeks later a meeting was held at Longbridge with Mr X and Head Office management and it was decided Mr X would return there in a new position. I subsequently learned that Mr X had a CV which showed he had moved from job to job in the company almost every six months. Now it would be some other manager's misfortune to get Mr X. Soon Tim Davies, who had been a Sales Supervisor of the region, took on the role of Franchising Coordinator and I was pleased I now had a much more capable and trustworthy colleague.

On 11th February 1976 I attended the Stamford Magistrates Court as the result of a driving offence that I decided to contest. In December 1975 I had a meeting at Longbridge which required me to leave home very early in the morning. Stamford, a town in Lincolnshire, had a number of narrow roads and on leaving the town the road I was on had one of those narrow points where there was a 'give way to oncoming vehicles' sign. The sign indicated I had to give priority to oncoming traffic and there were dotted line markings on the road to show this.

However, it was still very dark so I could see there were no vehicles approaching, and although I slowed down I did not stop but continued at a fairly slow speed. A mile or so further on I saw a police car's lights flashing behind me and after overtaking me it brought me to a stop. I was told I would be charged with not stopping at a give-way sign – even though I made it very clear to the police there were no vehicles coming the other way.

I was quite angry at this charge as I thought it was unjustified, and I decided to seek legal advice through the Automobile Association (the AA). There was free legal advice to members at that time and I was told I had a very good case as the road traffic act permitted the 'trickling' of a vehicle across a give-way sign in the hours of darkness if it was clear to do so. 'Trickling' apparently meant to move very slowly. 'Where there is a Give Way sign you must give way to traffic, but if on approach you can clearly see it is safe to emerge then you can do so without stopping.'

The AA said it would provide a lawyer to defend me and so I agreed to fight the charge. On 11th February 1976 I attended the Stamford Magistrates Court. My case was not the first to be heard and I sat and listened to one involving a violent attack on a person and a few other serious offences.

When it came to my case a policeman read out details of my alleged offence and he was followed by my lawyer, who cited the section of the Road Traffic Act in my defence. There were three very elderly magistrates who briefly conferred and then got up and left the court. My lawyer turned round as I was behind him and just gave me a thumbs-up sign. The magistrates soon returned to say that it was a dangerous junction and they could not let me off the charge as it would set a serious precedent. I was fined £20 and received three points on my driving licence. My reaction was that the magistrates had made the wrong decision and that the so-called precedent, if made, would reduce the level of income from fines issued.

In the last half of 1975 and in the early months of 1976 I attended many Franchise Policy meetings at Longbridge, in order to decide how the integration of the BL franchises and its implementation would take place in practice. In June 1976 most of the sales and marketing Head Office management relocated to new offices in Redditch, in Worcestershire, and so my meetings were now held there, still a long cross-country drive from home in Market Deeping.

It was decided by top management that all distributors and retail dealers representing BL Cars would hold a single core franchise which included all Austin Morris and Rover Triumph vehicles. This meant there would be a complete integration of these brands. It was also decided that the specialist brands of Jaguar Daimler, Land Rover and Sherpa vans, with their relatively much lower sales volumes, would be selectively awarded as franchises. However, the outlets receiving one or more of these specialist brands would have to operate the main Austin Morris Rover Triumph franchise, and then have the capability and resources in order to be considered for these extra specialist franchises.

It was clear to me that where there was one existing Austin Morris distributor and one Rover Triumph distributor in a city or large town it was most likely only one would be required and so selected to represent the one integrated Austin Morris Rover Triumph franchise in the future. Similarly where there was one existing Austin Morris dealer and one Rover Triumph dealer in a town, again it was most likely only one dealer would be required. These two scenarios would be the typical franchising situations we would face and would in practice lead to a further round of dealer terminations and a much-reduced dealer network.

I remember there was a great deal of speculation, anticipation and concern in every distributor and dealer, whether big or small. All of them naturally wanted to know as soon as possible what their future was going to be in the new BL Cars era. The Rover Triumph outlets were particularly fearful of their fate at the hands of what they considered had been an Austin Morris takeover.

I attended a number of premises planning meetings to discuss in the future dealer facility requirements of the new franchise arrangements as well as many corporate identity meetings to decide how the franchises and brands would be marketed and featured on dealer premises.

In anticipation of all this I started to get very involved in developing franchise plans for the region. For each existing Austin Morris distribution territory, I had to calculate the future sales potential based on company sales forecasts and then determine the optimum distribution and dealer network. My franchise planning work at the desk then had to indicate how best to use and develop the existing network to meet my optimum planning.

At that time a number of large national and regional dealer groups such as Dutton Forshaw, Henlys, Kennings, Lex, Mann Egerton, Marshalls and Wadham Stringer often operated BL franchises in the same main locations and clearly there would have to be an overall plan for the region that did not greatly benefit one, or greatly disadvantage or destroy another. Where the dealer group represented BL across more than one region it was accepted there would need to be recognition and discussion of the proposed franchising decisions on a national basis.

In addition, in the Rover Triumph and Jaguar Daimler networks there were high-quality companies that were now faced with the prospect of having to sell and service Austin Morris vehicles in order to retain the specialist franchises they held. I know many of them were not very keen on that prospect, and would have to decide whether they could demonstrate to people like me their commitment to the volume car side of the business.

Some family businesses did not want to get involved in what they called 'metal moving' Austin Morris and they also realized they were not likely to be the 'preferred candidate' in their market, where for example there were dealer groups. In some of these cases I was approached to see if we would support a negotiated sale of their business with the specialist Rover Triumph or Jaguar Daimler to the preferred candidate. In most circumstances this was the best outcome as it would ensure continuity of the business, and maintain staff employment and customer loyalty. It would also avoid any negative news in the local press and business disruption that could arise from the termination of an existing company and the appointment of a new franchise holder.

However, there were many locations where we could not engineer a negotiated arrangement of this sort. Ideally we would be looking for one distributor to represent all the brands but there were often three distributor incumbents all looking to be accommodated in the future franchise plans. It was clear we had to resolve if possible the distributor arrangements before we could confirm our franchise intentions for the retail dealers. After much work and debate within the company, the proposed franchise plans were finalized and we now had to go and present them to the distributor network.

Some of the first dealer group meetings were held in the last half of 1976. I recall a meeting in September with the Wadham Stringer Motor Group in London and I joined Bernard Bates and Bob Blackshaw to present our plans. A similar meeting took place with the Marshall Motor Group at Birmingham Airport, attended once again by Bernard, Bob and me.

From time to time we would see new vehicles introduced. A very significant new car came in April 1976 – I attended a Rover 3500 launch at Longbridge. The car became more widely known as the Rover SD1. 'SD' referred to the Specialist Division and the '1' referred to the fact it was the first car to come from the in-house design team. The car was launched with just a V8 engine, but it was well received by the press and it won the 1977 European Car of the Year award.

More models were to be introduced and I attended the launch of the Rover 2300 and 2600 in London in September 1977. These were smaller six-cylinder-engine versions of the car. The Rover SD1 was the final Rover-badged vehicle to be produced at Solihull, the last car designed largely by ex-Rover Company engineers and also the final Rover car to be fitted with the Rover V8 engine. Even though a number of vehicles would later be called Rovers, one could say the SD1 was the last 'true' Rover. Although this model had a number of quality issues I would later have two Rover 2600s as company cars which I enjoyed driving.

In my diary for 1976 I made some brief references to events that were not totally work-related. The summer of 1976 was one of the hottest on record and I remember a meeting with a Director of a large dealer group which culminated with a lunch at a local hotel. I was accompanied by Roger Charlton and the conversation turned to the subject of the hot weather, and the dealer group Director said he was pleased to have a swimming pool. He asked if I had one and I said I had a very small one. It was not a total lie as we had a paddling pool kept full of water in the garden for our daughter Gilly. The conversation moved on to other leisure pursuits and we were informed that one of his was horse-riding and that he had been taught to ride by his wife. I'm sure it was a harmless comment except I could hardly contain my laughter when Roger spontaneously said something that implied a double meaning. Not for repeating here. I don't think the Director spotted this and continued his remarks on the subject as if nothing untoward had happened.

Over the next few years a number of sporting events were arranged between the BL Cars Zone 3 team and staff from the Mann Egerton dealer group. I recall the games were meant to help develop good working relationships between the two companies during a period of network reorganization. Also the matches would be management against management as we feared heavy defeats if Mann Egerton included young active sporting staff from its dealerships.

The first of these matches took place in October 1976 when the BL Cars Zone 3 Regional Office football team played against the Mann Egerton dealer group team and according to my diary Zone 3 won 6-2. The following May we had a return match against Mann Egerton one Friday evening in King's Lynn and my diary shows we won 2-1. I don't recall much about these two matches other than Roger Charlton was our goalkeeper and at around 6 foot 5 inches tall it was surprising that a penalty was scored against him in one of these games with the ball being lobbed over his head.

My diary for 1977 has numerous references to internal franchise strategy meetings, held at Longbridge HQ and the Regional Office. In March one diary entry says 'franchise strategy meeting – final zone sign off' and I know that a large number of dealer group meetings subsequently took place through the summer months.

One such meeting in July was with the Mann Egerton Group at its Norwich Head Office. Bernard Bates and Bob Blackshaw flew in from Birmingham to join me and David Evans, my opposite number from the London/South East Region. There we met Jim Campbell, the Chairman, and his Regional Directors Sidney Harris, Brian Bleaney, Archie Clayton and Jim Clifton.

Most of the Mann Egerton branches were in the Zone 3 Region and the franchise proposals I presented did confirm franchise continuity and growth opportunities in most locations. However there was some adverse impact on Mann Egerton with some of its outlets not being 'preferred candidates'. At the start of the meeting Jim Campbell made their position clear by stating that his company had in recent times divested itself of many Ford outlets in favour of BL and it would therefore not accept any proposed reduction in its BL representation. What I had to say was not welcomed and discussions between us continued for many months afterwards until a location-by-location agreement was eventually achieved.

Once again these nonpreferred outlets would in due course take on competitors' franchises. The franchise planning called for the long-term candidates to pick up the sales of outgoing distributors and dealers in order to maintain market share. However this was unrealistic as the outgoing outlets were high-sales-volume outlets and would convert most of their customers to the products of their new franchises. All this would have a serious adverse impact on BL Cars's sales and market share.

There were times when a dealer visit did leave a lasting impression on me. One of these came early in 1978 when I had to visit a small dealer in the village of Leverton, a few miles north of Boston in Lincolnshire, to confirm the termination of its franchise. I mention this visit because for me it was unique to terminate a dealership owned and run by someone who had not only won four Manx GP TT motorcycle races on the Isle of Man, but a man who did it with only one eye. The person was Austin Munks, who suffered a terrible shooting accident in October 1935. While he was out on an organized shoot one stray pellet hit him in his left eye, causing him to lose his sight in that eye. However, the following year he went on to become the first double Manx Grand Prix winner, which was a remarkable achievement.

In January 1978 the Zone 3 Regional Office relocated from Lincoln Road, Peterborough to brand-new offices at Apex House on Oundle Road, Peterborough. This move, which was organized by Ken Smith, the Zone 3 General Manager, gave us all a much better working environment than we had before. However, Leyland Cars was still in trouble financially and as we started the year 1978 we were all waiting to hear what the new top management was going to do. We didn't have to wait very long.

A major conference in the company was held on 3rd February 1978 at which Michael Edwardes, a South African businessman who had been appointed Chairman and Managing Director, addressed the UK distributors and dealers at the Wembley Conference Centre. Regional management in the UK was also invited to hear what he had to say about the current state of the company and his plans for it. Much has been written about Michael Edwardes and what he did during his time at BL so I will make just a few comments about what was going on at that time in BL.

In January 1976, Edwardes had joined the National Enterprise Board (NEB) which was running BL. By then BL was suffering from severe disruption through industrial action and the Labour government made it clear it was not prepared to continue providing financial support for BL. Strikes continued across the company and in 1976 alone there were serious stoppages at Browns Lane (Jaguar), Canley (Triumph), Cowley (Morris), Drews Lane (vans), Llanelli (radiators), Longbridge (Austin) and Solihull (Rover). In the first five months of 1976 it was estimated that these strikes cost the company around 60,000 cars in lost production, and in 1977 as a whole strikes cost the company around 250,000 cars in lost production. The union leaders no doubt felt they were representing the best interests of their members but they were in fact seriously crippling the company.

At Longbridge, the man who had led the workers to many strikes was Derek Robinson, the union convenor there. 'Red Robbo', as he was known in the national press, was responsible for bringing to a halt the production line of the largest factory in BL, and this brought him into direct conflict with Edwardes. To return Longbridge to normal production Edwardes decided Red Robbo had to go, but this didn't happen until November 1979.

The government began to wonder whether BL would survive and realized the Ryder Plan wasn't working. The toolmakers' strike which went on for four months was one of the most damaging of all the strikes that occurred in BL, and as a result the BL Board decided it would not consider asking the government for more financial aid until the strike was broken. However, when the strike was over BL management had to ask the government for a further £100 million.

Edwardes had been brought in to provide the leadership that had been lacking. He spoke of management's right to manage and that he had to make big changes in the company in order for it to survive. Change meant improving the products and model ranges on offer to customers and 'sorting out' the unions in order to increase productivity. Initially what he faced was another cash crisis, which was resolved with more funds being provided by the banks and the government.

In February 1978 he presented his plans to save the company to a delegation of shop stewards and union and employee representatives. He told them he needed to cut out the excesses and waste in the company in order to save it. He received an almost unanimous vote of support in favour of his plans. The same month BL managers were invited to a conference at Wembley and we heard a similar story direct from him.

Although he was small in height, Michael Edwardes came across to me as a very strong, charismatic and determined character, who knew what needed doing and also what he was going to do. He spoke directly to us all with just a few notes and not the usual presentation from a lectern. I think we were all greatly encouraged by what he said and I left the Wembley Conference Centre with a much more positive feeling about BL and its prospects for the future. It certainly needed someone with a strong personality to take on the unions, because BL had become something of a laughing-stock and an object of much ridicule among the general public.

Throughout the first half of 1978 I had meetings with a large number of distributors and dealers aimed at securing commitments from them to invest in further developing our franchises. None of this was easy as sales of the company's cars were still declining.

Chapter 6

# Territorial Army Exercise 1978

**ON FRIDAY, 21ST JULY 1978** my friend and work colleague Roger Charlton phoned to ask if I would like to join him on a fun day in Norfolk and explained that Neil Johnson, one of our Directors, was looking for some of us to take part in a Territorial Army 'exercise'. Neil Johnson was not only the Service Director in BL but also a Lieutenant Colonel in the Royal Green Jackets, Territorial Army. I said yes to Roger's invitation without giving it much thought or getting any more information from him. It would turn out to be a most memorable activity.

Before recounting the events that followed, I want to say something about the Territorial Army, or TA for short. The TA (now known as the Army Reserve) is the active-duty volunteer reserve force and an integrated element of the British Army. During periods of war, the Army Reserve is incorporated into the Regular Army for the duration of hostilities or until deactivation is decided. Army Reservists normally have a full-time civilian job, which can provide skills and expertise directly transferable to a specialist military role, such as NHS employees serving in Reservist Army Medical Services units. All Army Reservists have their civilian jobs protected by law should they be compulsorily mobilized.

I will now explain what happened. On the morning of Saturday, 22nd July Roger Charlton and I drove together to the meeting point, a pub near Thetford in Norfolk. It was a warm day at home and in Norfolk and we had come to Thetford just in the clothes we wore, which in my case were light-blue denim jeans and a light roll-neck sweater. We arrived at the pub around 1 p.m. in time for some lunch. There we met our colleague John Goodyear, a senior manager in the Leyland Cars Service Division at Cowley, Oxford, who reported to Neil Johnson, and who had been charged by Neil with obtaining some BL managers for the TA exercise. John informed us that we and some friends from his team at Cowley would become a small group of just eight mercenaries who would be hunted down by the TA that weekend. Roger and I were taken aback by the word 'weekend' as we thought whatever we were going to do there was just on that Saturday. In addition, given the words 'hunted down', it didn't sound like the fun that Roger had mentioned to me.

At that point Roger and I went to a nearby phone box to phone home with the news about it being a weekend event. I phoned my wife Annette to say I would not be coming home that evening after all and I passed on the little information I had been given about what we were going to do. The eight of us comprised Roger Charlton, John Goodyear, Malcolm Smith, Roger Storey, Brendan, Danny, Keith and me. Shortly after making my phone call home the eight of us were picked up in a truck and taken inside an Army battle area where we were told we would be briefed on the Weekend Exercise that would take place there. I didn't know anything about the battle area at the time. Here are some words about it.

The Stanford Battle Area is a British Army training area situated in the English county of Norfolk. The area is around 30,000 acres (120 km$^2$) in size, and is situated 7 miles north of the town of Thetford and 25 miles south-west of the city of Norwich. The area was originally established in 1942 when a battle training area was required, and within it a 'Nazi village' was created. The whole area was used in advance of the D-Day invasion of France. This involved the evacuation of the villages of Buckenham Tofts, Langford, Stanford, Sturston, Tottington and West Tofts. It is now known as the Stanford Training Area and as it is a live-firing area, access is not allowed without special permission from the Army.

After the short journey in the truck we were taken into a large hut where a senior officer in the regular British Army briefed us. There was no sign of Neil Johnson! We were informed that we would be taken to an area in West Tofts, which comprised some buildings, most of which were derelict, surrounded by many acres of forest, bush and tall grass. The building in the middle of this vast area would effectively be our mercenaries' hideout or base camp.

We were told that for the rest of Saturday afternoon we would be trained in how to use an SLR (self-loading rifle), ammunition and the thunderflashes (very noisy pyrotechnic devices like grenades used especially in military exercises). After this we would get a meal and drinks before being taken to our 'base' to prepare our defence. The Exercise would commence at exactly 7 p.m. and our objective would be to avoid capture by exactly 5 a.m. on the Sunday.

We were then informed we would have to sign a disclaimer in the event we were injured or worse, and then taken outside to receive our 'training'.

We were shown how to use an SLR and the ammunition and the thunderflashes. I remember asking the Army instructor about the dangers of 'blanks' as ammunition, and to answer this he fired a blank directly into a sack of hay at point-blank range, and it blew a big hole in it. It was now very clear to me and the others that the so-called 'fun day' that Roger had suggested was not going to be fun.

We were then informed that the TA on this exercise would number around eighty men. We were told we would have the benefit of knowing the size of our enemy but they would not know how many of us mercenaries there were. I couldn't see how this could be a benefit as we would be greatly outnumbered! A serious final message I recall to us from the senior officer on this 'exercise' was 'treat it seriously, because they will'!

Later in the afternoon we had the meal we had been promised and realized this would be the last one until the exercise was over. I wondered what was in store for us as we were the enemy of the British Army and we were going to be hunted and attacked by it as if it was a real 'mission' and we were a real enemy.

We arrived at our base camp around 5 p.m. I could see there were many buildings, but most of them were derelict. The biggest building looked the least damaged and we agreed this would be our headquarters or base camp. What struck me most was that we seemed to be surrounded by many acres of forest, tall bushes and tall, thick grass. The map we had been given supported the fact we were indeed in a heavily wooded area where each of us could easily get lost. In which case would the British Army easily find us?

The main building was oblong in shape, measuring around 32 feet by 15 feet. Inside there were two walls which almost separated the building into three rooms. The two outer 'rooms' were approximately 8 feet by 15 feet and had doorways, but no doors, so we had two entrances. Each of these rooms also had windows, but none had any glass. The inner 'room' was square-shaped and the largest, measuring 15 feet by 15 feet. It had two large windows but again without any glass. There was no furniture in the whole building.

Once we had given the main building and the immediate area around it a very brief survey we agreed we needed to decide upon a plan to avoid capture during the next twelve hours. That was our main objective. One of the main problems we had was not having any idea when or how we would be attacked. So we decided we should simply try to defend ourselves the best way we could. Several options were discussed.

One of these was for all of us to stay in the building and have our equivalent of Custer's Last Stand. An alternative option was to divide ourselves into four teams of two and for each duo literally to go and hide somewhere in the forest or vast expanse of tall grass and hope not to be found, a military version of the game hide-and-seek.

We decided to explore the hide-and-seek option first of all with each duo going out to explore possible places to hide but we agreed that at worst the fall-back position would be for each duo to return to our headquarters where they would await an inevitable attack. As this was a likely outcome we decided to defend our building as best we could.

Someone suggested setting up tripwires on several well-trodden routes to our building and not thinking about the consequences we set up a number of these. We were now almost instinctively starting to think as you would do if you were being hunted. We were treating the exercise seriously. Each of us had been given an SLR, a magazine of twenty rounds of ammunition plus another twenty rounds. We also had been given about a dozen thunderflashes between us. We had no other equipment we could use to defend ourselves.

However, a couple of our group of mercenaries hadn't started to take it seriously. They decided they would go into the nearest village for some liquid refreshments having calculated there was enough time to do so before the 7 p.m. start time. I can't remember who went but the liquid they brought back was alcoholic.

In late July we knew sunset would not be until 9 p.m. and sunrise at about 5 a.m. so we would have around eight hours of darkness. In our camp we wondered whether the TA would attack us in the next few hours of the evening while it was light, or during the night. We didn't know. Yet very soon we thought we knew. At 7 p.m. we suddenly heard the unmistakable sounds of helicopters in the vicinity. We couldn't see them because of the trees that surrounded our camp, but they didn't seem to be very far away and there seemed to be many of them – enough for eighty soldiers. Soon the noise of the helicopters stopped. They have landed! But where?

I think the sound of helicopters sent our anxiety to fairly high levels. I know it did mine. Our unanimous decision in our state of mild panic was to go out in teams of two and hide. We had not agreed where we would go and how we would communicate once we did this. Roger and I took our SLRs and made for the large forest. On entering, I was suddenly acutely aware of even the slightest sound, mainly because I sensed that even a small sound could be a TA soldier looking for me. I was also very conscious of not trying to make a sound. After a while I started to feel very warm. It had been a warm day but now deep in the forest, thick with trees, it seemed even warmer.

It is interesting what your imagination does to you in a potentially threatening and dangerous situation, and one that you have never experienced before. I was expecting to be confronted by a TA soldier at any moment. Would I fire at the soldier immediately? Would I run? Would Roger and I stay together and fight? In addition to hearing even the smallest sound I now started to see even the slightest movement in the trees and undergrowth. My senses of sight and sound had become even sharper and I could tell my heart rate was higher than normal.

I don't know how long we had been there, maybe an hour, when suddenly around 20 yards away through the trees I was sure I saw the top of a head moving slowly behind a stone wall. I immediately fired at the head although I was not sure what good it was going to do. The person behind the wall turned towards me. It was one of our mercenary team! If I had thought for a moment I should have realized if it had been a TA soldier he would have had an army beret on his head. Much more significantly by firing instinctively I also realized I had given the TA an indication of our exact position.

Roger and I now moved out of the trees towards a vast expanse of tall, thick grass. Once we reached the middle of this we sat down and pondered what to do next. For some time we just stayed where we were thinking this was perhaps a good place to hide. However, when it was starting to get dark we thought this was not where we wanted to be for perhaps the next eight hours, and we slowly made our way back to the main building. As we approached we could just hear some voices and so being very cautious we entered the building. It was still sufficiently light to see the other six members of our team. They too thought this was a better place to spend the night than in the forest or in the long grass.

It was agreed we would all see out the exercise here, and carry out our version of Custer's Last Stand. We decided four of us would keep watch for a couple of hours, and then the other four would take over. We thought that was all we could do through the night and this might allow us to get a few hours' sleep. We had lasted two hours, we had just eight more to go.

In practice, our planned lookout duties didn't last. One of the team heard a noise outside and we all cautiously walked out. It was very dark and I certainly couldn't see anything. However, on the spur of the moment one of the guys decided to solve this problem by setting off a thunderflash. In the circumstances this was not a very clever thing to do. Yes, it did create a great deal of bright light for a short while, but as I indicated earlier a thunderflash when set off is incredibly noisy. We gained nothing because we saw nothing, but the TA must have known now where we were if they didn't already.

We all went inside. We were all awake and we all now expected the inevitable – an imminent attack. However it didn't come. One hour passed, and then another. It was now in the early hours of Sunday, and we were all tired, thirsty and I was sure the attack must come any moment. But it didn't. As time passed we were still all crouched by a window looking out into darkness, seeing things but not seeing soldiers.

We became convinced we were being encircled by the TA and that any moment the soldiers would attack. Personally I didn't know what form this attack would take. However we would unleash our SLRs and thunderflashes so I expected the same from them. It wouldn't be pleasant. As each minute and hour passed the expectation that a dramatic end is very near raised our anxiety, heart rate and fear to very high levels.

At around 4 a.m. and with only an hour to go to achieve our objective a strange atmosphere and conversation emerged in our group of eight. Maybe we could avoid capture by 5 a.m. None of us had slept during the night and with all the adrenalin we had used up we were all pretty much worn-out, but not as fearful of the outcome as we were at 7 p.m. the previous evening or even just a few hours earlier. I was now looking at my watch every few minutes. We were still here. Where were they?

It was starting to get light at 4.40 a.m. and still they hadn't come to attack us. Five minutes later nothing had changed. We had been on guard all night waiting for an attack that hadn't come. At 4:50 a.m. we were all starting to think we might do it. I couldn't understand what was going on. It was lighter outside now. I could look out of the window towards the tall, thick grass that stretched over a hundred yards into the distance. I couldn't see or hear anything. At 4:55 a.m. we only had to survive another five minutes. We were going to achieve our objective, weren't we?

And then I saw them at the far end of the grass field, running sideways and forwards like hundreds of ants, but getting bigger and closer every second. A group of ants is sometimes called an army. Well, the army was now here.

In no time at all we started firing and throwing thunderflashes from where we were. The same was coming back at us, but much more of it. Suddenly our building was full of soldiers and we were being physically attacked. The noise of the thunderflashes inside our building together with the loud shouting and screaming of the soldiers was incredibly deafening. The room was immediately full of smoke. Before I could do anything I was lifted and dragged outside by a couple of soldiers and literally dumped on the ground alongside my colleagues. It was all over in probably less than two minutes.

I can recall a few of my colleagues shouting 'We surrender', and 'Don't hurt us', but that didn't stop the TA soldiers doing what they had been trained to do. We were thoroughly 'roughed up' as if we were the enemy in reality and they had treated it seriously. Fortunately and to my great relief while I was hurt I wasn't injured, and I think that also applied to my seven colleagues. But we were all shaken by the experience and the violent ending. A photograph was taken shortly after showing the eight of us as prisoners with our hands in the air, indicating our surrender. I have that photo as well as one taken just before our adventure started.

Neil Johnson, the TA's commander, then appeared to see if we were all OK and he thanked us for contributing to a successful exercise. It seems our trip wires had some effect as during the TA's attack a sergeant had slightly injured himself running into one of them. I don't recall any of us being sympathetic or saying sorry. We weren't sorry!

After a short while Neil surprised us by saying that this wasn't the end of the exercise for us. He told us that the TA soldiers would soon start a march back to their base, which would involve them crossing a bridge over a river. The eight of us would be taken to the bridge without their knowledge and we would hide under the bridge and await the TA. As soon as they got onto the bridge we would attack them.

The idea of us getting our own back on the TA sounded good but it then occurred to me the one big fight we had just had with the TA was enough for me. How would the TA react and could we expect to be 'roughed up' again? However, we were told we had no choice but to go ahead with what was now part two of the Neil Johnson 'exercise'. Apparently there would be no 'fight' and what we 'mercenaries' were doing was for the TA leadership to see how their men would react to an unexpected attack on them.

I don't remember much about what happened except for that I was much calmer about what was going to happen. I think that was because I had used up all my energy and there was also for some reason no anxiety left. Even as the TA marched in formation towards the bridge, we were just waiting to fire at them. We waited until most of them were on the bridge and 'attacked'. The surprise worked well because many of the soldiers just jumped off the bridge into the river. There was no fight this time and again it was all over in a minute or two. However, this time there was some acknowledgement from the soldiers of our courage in being willing to attack them, especially as there were only eight of us.

It was still very early in the morning when we all gathered together for a breakfast outside. It had been a very long twelve hours. There was some interesting information that Neil Johnson gave us at the end of the exercise over breakfast. When the TA soldiers landed in their helicopters at around 7 p.m. the previous evening most of them camped down and slept through the night. A small number of soldiers were tasked with reconnaissance, to determine the exact number of mercenaries and their exact location. Apparently it was not too long before they had all the information they needed and which enabled them to plan and achieve their mission by 5 a.m., which they did. I have to say it was timed to perfection.

Writing about it all these years later I was surprised how much I remember. After the event I thought about writing a short book about it and drew a rough sketch of the mercenaries' building as well as the surrounding forest, bushes and tall grass. I also thought that this TA adventure had the possible makings of a good short film for television with the mercenaries appearing to be real and preparing for ten hours to avoid capture and then waiting to be attacked by the opposing British Army. Not until the mercenaries were captured and defeated would it be revealed it was a TA weekend exercise and then the early part of this story up to the point of the TA briefing could be shown.

The TA 'exercise' was one of my most memorable weekends. It gave me some understanding of what a British soldier does, and a little about what the enemy of the British Army must feel when being attacked. And I found out a little more about myself in a situation I had not come close to experiencing before. I'm very glad I did it.

Chapter 7

# British Leyland (BL Cars) 1978-1981

**IN AUGUST 1978 WE WERE** informed that senior management would be required to undertake some compulsory company psychological tests in London. It seemed Michael Edwardes had decided on this plan which most of us assumed was aimed at ensuring the right managers were in the right jobs. My appointment with Eric Jones, the person organizing these tests, was in Upper Berkeley Street, London at 9 a.m. on Monday, 4th September.

These tests lasted most of the day and included those that were designed to measure my intelligence, including numeracy and vocabulary. I remember a session where I had to make spontaneous decisions about which of two options I preferred. There were a large number of these 'preferences' and many of them seemed slightly ridiculous to me. One I remember was as follows: Would you prefer to have a sexual encounter with a member of the opposite sex or a tidy desk? With regard to these preferences I just went for those that appealed to me most as we had to complete the test within a fixed time period and this meant there was no time to think about the answer. I couldn't help thinking afterwards what strange persons thought up these preference examples.

After this day of tests I wondered whether the results would have any bearing on my career development. I wasn't bothered about the outcome as I thought I had done well in the test sessions, but I am sure there were many who were concerned.

On 22nd November 1978, Peter Johnson, Operations Director Car Sales, circulated an internal organization communication confirming my appointment as Field Sales Manager, Zone 3. I was informed of this good news by Ken Smith earlier in the month. I would continue to report to Ken, General Manager of the region, who on 1st October had replaced Chris Martin. I was taking over from my good friend and colleague Roger Charlton, who was going to a new position as Sales Promotions Manager, based at Redditch.

I had been carrying out the role of Franchise Development Manager for over three years, which was long enough, so I was delighted to get back into an operational sales job and a new challenge. I don't know whether the psychological tests I had taken had a significant bearing on my move into sales management, but there were many management and other organizational changes that took place around that time so perhaps they did.

The Field Sales Manager position was an important new job for me because I would now be managing a field force of Area Managers, and the role brought with it the responsibility for achieving the Zone 3 region's vehicle sales objectives. The company's continuing decline in vehicle sales would put more pressure on getting results than in my previous job.

In December I had a meeting with Alan Mazdon, the new Sales Director of Jaguar Rover Triumph (JRT), who had come from Renault. I don't remember what we talked about, but I had a telephone conversation with him less than a year later I will never forget.

So 1979 would turn out to be a year of big changes for me. However, the first event of any interest in my diary was the Jaguar Series III launch to the dealers I attended on 23rd February at the Jaguar factory in Browns Lane, Coventry. The public launch was on 28th March. I must say while it was the first Jaguar car launch I attended it was not in any way exciting and as a result I don't remember anything about it. This new vehicle was in fact just a facelift of the Jaguar Series II which was produced from 1973 until 1979. The Series II models were known for their poor build quality, which was attributed to Jaguar being part of BL as well as the trade union relations problems that plagued the company in the 1970s.

From my experience and with responsibility for selling the car, the Jaguar Series III suffered from poor build quality and there was for me an infamous period where the car was only built and available to customers in three solid-body colours, namely Damson Red, Cotswold Yellow and Tudor White. With minor changes and improvements this model ran until 1992.

In April, May and June of 1979 I had a number of meetings with Alan Mazdon, the Sales Director, and John Taylor, the Sales Administration Manager, regarding the setting up of a new JRT sales organization, which would operate separately from Austin Morris in the UK. I recall that the BL Cars structure continued but just for Austin Morris, while the JRT sales structure would have to be created from scratch. My meetings in June covered JRT financial budgets and major operational matters. In the one on 5th June Mazdon confirmed an offer for me to become General Sales Manager for the northern half of the UK.

At the end of May a major Austin Morris dealer conference was held at the Royal Lancaster Hotel in London. The main points and issues raised at the conference were summarized in a special edition of *Sales Talk*, the company's magazine which was distributed both internally and to the dealer network. I have a copy of this *Sales Talk* and so I can quote accurately from it about the conference.

Ray Horrocks, Chairman and Managing Director of Austin Morris, said, 'Austin Morris is in a state of continuous development.' This was an interesting statement in my opinion, as ever since I joined the company in 1969 it was in a state of continuous change, rather than development. He also spoke of the joint venture with Honda, which was the first time we heard about it. He said a new medium-range high-specification car developed by Honda would be built at JRT and sold in the EEC and Britain under the Triumph marque through the existing BL network. Production would start in about two years.

He also spoke of the plans to give Austin Morris and JRT separate responsibility for their own field sales operations. I was already involved in the JRT part but it was the first time this message was communicated to the distributor and dealer network. It would soon start alarm bells ringing particularly in the major dealer groups who foresaw an adverse effect of a potential split in two of the integrated AMRT franchise that had been created by the Ryder Plan only four years earlier.

Trevor Taylor, Director UK Operations of Austin Morris, spoke of changes in the Austin Morris organization and of concerns on market share and sales performance. He highlighted a market share of 16.6% in the first four months of 1979 – a shortfall of 2.5% against the task with virtually free supply stock. He said it was 'extremely concerning and disappointing' and he referred to a 'disastrous April'.

There were positive presentations from Peter Johnson, Sales Director, Harold Musgrove, Director of Manufacturing, and Tony Cumming, Director Marketing Operations, who all spoke about their plans and support being given to the network to promote vehicle sales. Trevor Taylor closed the conference with what *Sales Talk* called 'challenging words'. He said, 'Now is our chance to show what we are made of when the chips are down. We must attack the market as never before and prove that British is best.'

In my experience of attending many network conferences such as these, they were always upbeat and the presentations were as positive as they could be. This conference was no different but despite the positive message the fact was that Austin Morris products at that time such as the Austin Allegro and Morris Marina were far from being the best. In 1969 the Austin-Morris Division of BL had a market share in the UK of around 28%. By the end of 1979 and in just ten years despite everyone's best efforts in the company and in the dealer network the Austin Morris market share had fallen to 16%, and it didn't stop there.

On 8th June 1979 I received a letter confirming my appointment as General Sales Manager in JRT reporting to the UK Sales Director. This promotion to corporate grade 30 gave me a big increase in my salary and I would be based at the BL Regional Office in Altrincham, south of Manchester.

During June and July of 1979, I was effectively carrying out two roles. I was still Field Sales Manager in Zone 3 but also working in my JRT position to help set up the new JRT sales organization which was now scheduled to take effect from 1st August. The existing field force of Area Sales Managers (ASMs) was given a choice. Each could choose to stay with Austin Morris led by the existing top management of Trevor Taylor and Peter Johnson or take on a similar field sales role in JRT.

Nearly all the ASMs decided to go with Austin Morris which meant my colleague Mike Case and I found we had to create a new field sales team and this meant much interviewing during July. At the same time my new job meant a house and family move, and so early in July we put our house in Market Deeping up for sale.

On 1st August 1979 the new JRT sales organization came into being and held its first meeting at the company's Canley offices in Coventry. I remember this occasion well. I was excited by the new challenge but I also wondered how the split with Austin Morris would work out as there was a distinct possibility JRT would wish to develop an exclusive dealer network. The meeting continued the following day and in addition to discussing business plans and objectives all members of the field sales team collected their new Rover SD1 cars. Mine was a Triton Green Rover 2600, registration number GWK 609V.

On the weekend of 4th and 5th of August Annette and I went house-hunting in Cheshire and found a new house we liked being built at Wistaston, a village 3 miles north of Nantwich. We knew this area quite well as we had been living in Sandbach less than five years earlier. For me it seemed a little strange to be going back to the Cheshire area to live. During the following week I had to travel to Altrincham to discuss arrangements for my JRT team to be housed in the Zone 2 Regional Office where Austin Morris was the predominant presence. All seemed to be going well in the first few weeks of the new JRT sales organization.

However, nothing was to prepare me for what happened on 29th August. That day I went to Hunstanton on the north-Norfolk coast to present a cup to the winners of the Rover Golf Regional Final. I had just done this when a man came from the clubhouse to say I needed to take an urgent phone call from my boss.

At the end of the phone was Alan Mazdon and his first words were 'It's all off'. I didn't understand what he meant until he said that Michael Edwardes had been strongly lobbied by the Austin Morris distributors over their concerns for the viability of the Austin Morris franchise without the specialist cars of JRT. Quite simply Edwardes had decided to put the two sales organizations back together again with immediate effect. Mazdon added that this meant the Austin Morris sales structure would immediately become BL Cars again and that immediately members of the JRT sales team were now redundant. Mazdon said he would be leaving immediately and said I would have to phone everyone in the JRT team with this shocking news. I asked him, 'What do I tell them?' I shall always remember what he told me. He simply said, 'You will think of something.' That was my last conversation with Alan Mazdon.

You can probably imagine my reaction to this. I was angry! Without any warning I was out of a job. I had sold my house in Market Deeping. I had a deposit on a house in Cheshire I no longer needed. I had to phone everyone in my team with bad news and to say each was redundant. I had to find some appropriate words. And what do I tell Annette my wife? All this went through my mind as I made the lonely journey back home.

When I arrived home, Annette said before you do anything ring Peter Johnson, who I assumed was now Sales Director BL Cars. I phoned Peter. He explained what was happening and said I would be offered the job of Field Sales Manager in Zone 5, the London and South-East region. I was told to phone my field managers and tell them to stay at home and await further news.

Among those managers we had recruited into JRT was Roger Knight.

I rang him first. It was one of the most difficult phone calls I have ever had to make. He had given up a good job with Volvo to join JRT. Now barely a month later I had to tell him JRT was no more and he had no job. There is some irony in this, because not long after he became a Regional Director of the Henlys Motor Group and this led him to the Marshall Motor Group, where in due course he became Managing Director. On the few occasions I have met Roger since then I said if it wasn't for me and the JRT saga he wouldn't have become MD of the Marshall Motor Group!

The day after the 'Hunstanton' phone call Annette and I joined other senior managers in hosting the winners of a dealer incentive campaign on a four-day trip to Rome. What a week that was!

September started with a number of organizational briefings in the company and on 13th another network conference took place at the Wembley Conference Centre to inform the network once again on the company's new situation and its future plans.

We started house-hunting again, this time in Surrey. The Zone 5 Regional Office was situated at 41–46 Piccadilly in the centre of London, so finding a good house we could afford within 30 miles of the office was not easy. The Piccadilly Office was also the BL Headquarters where Michael Edwardes was based.

I started my new job on 1st October 1979 reporting to Graham Powell, General Manager of the region. I was immediately involved in developing the 1980 Business Plan for the region and holding initial review meetings with my team of ASMs that comprised Jack Beattie, Jim Campbell, Dave Cooper, Ian Goudie, Simon Gow, Tim Lewis, John Meaden, Graham Osborne, Ian Tagholm and Jim Wright.

After much searching, on 27th October we finally found a new house being built on the southern side of Camberley in Surrey. Leaving home at 7 a.m. latest I could get to the office by 8 and into the underground car park where I had one of the few car parking spaces. However, until our new house was finished I would have to travel from home in Market Deeping by train from Peterborough to London and for most of the week I was staying overnight in London hotels. It was not what I wanted but commuting by car from Market Deeping into central London was not a viable or sensible travel option.

BL Cars's sales performance in the London area was nearly always below the national average market share and below the sales targets that had been set by top management, so there was added pressure on Graham Powell, me and the ASMs in what was the largest car market of all the six regions. In addition, it was becoming increasingly fashionable for customers, especially in London and the South-East, to buy imported cars from European and Japanese motor manufacturers, and Ford was increasingly dominant in the fleet sector.

With continued pressures from Head Office, I soon found myself spending much of my time in meetings with the management of Dorlands, our advertising agency, at their offices in London. I had to manage and make the best use of a very large marketing and adverting budget and my meetings with Dorlands were usually weekly events to decide on the content of advertising which we placed almost daily in the *Evening News*, the *Evening Standard* as well as on LBC Radio and Capital Radio. In addition, and in conjunction with Dorlands, we produced and sent regular marketing communications to the Zone 5 distributor and dealer network, in a slim brochure called *Zone 5 Adnews*. The objective of *Adnews* was to clearly inform the network of the latest sales incentive campaigns and details of marketing support.

One incident I remember at the London Regional Office involved the BL Chairman Michael Edwardes, who also had his office in the same Piccadilly building. At the time he was frequently giving interviews to the national press on the financial state of the company and issues with the trade unions. One day I received an internal phone call to say he wanted to use my office for a press interview and that he was on his way to my office 'now'. I barely had time to do a bit of tidying up before he entered. He briefly acknowledged me before I made a quick exit. My office, being fairly small compared to others in the building, seemed an inappropriate place to conduct an interview, but that's what happened.

In the last quarter of 1979 wholesaling the required vehicle allocation within the region continued to be difficult. Our monthly wholesale meeting had one additional element. There were no parking spaces for the ASMs at the Piccadilly office so they parked outside in Sackville Street where each space had a parking meter. So the wholesale meeting was continually interrupted by one or more of the ASMs having to go out to 'feed' the meter with money before the time expired.

I remember the annual sales target in the London region was over 100,000 vehicles a year and with the network required to stock at least 12.5% of its annual target, at least 12,500 vehicles had to be stocked at any one time. However, as a result of the overall underperformance, the distributors in the region usually had more than adequate stocks of vehicles and vehicles on order without the need to order the additional vehicles we were trying to wholesale to them. The monthly wholesale negotiations became increasingly difficult and I recall having to be personally involved with the distributor groups, such as Henlys. In addition, and especially in Greater London, there was an increasing problem for the distributor network to find the physical space required to park and store the very large numbers of cars.

An example of the sales letter I sent out to the region's distributors was sent back to me many years later by Jim Lyall, Managing Director of Ray Powell Ltd, one of BL's largest distributors based at Ilford, Essex. Dated 19th November 1979 my letter states that 'we are seeing no improvement this month over the unacceptably low October performance'. It adds, 'Please ensure you are taking full advantage of the support we are giving.' I then refer to financial bonuses on offer and I end the letter with, 'Please keep the pressure on – we cannot afford to reduce our efforts over the remaining weeks of the year.'

Jim also sent back to me the telex/telegram I sent out to all dealer group heads also on 19th November confirming the main points of my sales letter. Looking at these communications we were wondering what else we could do in addition to the vast sums of money we were spending on advertising and sales bonuses we were making available to the network. We were continually urging more effort from our distributors and dealers. I think they were doing the best they could, but in the end we didn't have the vehicles desired by retail and fleet customers to deliver the sales we wanted. On 20th December 1979 we presented details of a 1980 sales campaign to the distributors. We would continue with more of the same, and it seemed sales campaigns were never-ending.

In December 1979 Jack Beattie, one of my ASMs, sadly died – and on Christmas Eve. I and all his colleagues from the London Regional Office attended his funeral at the Oxford Crematorium. It was bad way to end the year! January 1980 was also a very sad month because within the space of a few weeks I attended another funeral. This time on 9th January it was at the Plaistow Crematorium in East London, for the funeral of Kevin Sparrow, my cousin and son of my Uncle Roy. Kevin, who was just thirty years old, had tragically died of choking in his sleep after a heavy bout of drinking.

On Monday, 21st January 1980 we moved out of our home in Market Deeping. However our new house was not yet ready for us to move into so we started a stay at the Frimley Hall Hotel, Camberley, that would last five weeks until we finally moved into our new home on 26th February. This extended stay in a hotel meant we had to put our furniture into storage, Pippin our cat into a cattery and Sammy our dog into a kennel. Staying in a hotel for five weeks was a difficult time for Annette and Gilly as it totally disrupted their normal day-to-day routine of home and school life.

One aspect of my job was to make presentations from time to time to the distributor/dealer network in the region on sales-related subjects. On 15th February I made a presentation in one of the most prestigious venues, the British Academy of Film and Television Arts, better known as BAFTA, in Piccadilly, London. The presentation was memorable because making it on stage I could not see anyone of the hundred-plus audience in the theatre in front of me. It was dark apart from lighting to highlight me on stage.

I had one additional responsibility in my job. On the ground floor of the Piccadilly building looking out onto Piccadilly itself was a large car showroom capable of displaying around ten vehicles. One objective here was to promote the sale of BL's models to the tax-free sales and diplomatic sales sectors. My responsibility was to rotate the vehicles on display every month or so and to replace the vehicles with a fresh line-up of cars.

Normally these new cars would be offered and taken by dealers in London, but during the first few months of 1980 no dealer wanted them. They already had enough cars in stock and didn't want any more. So I decided I would use these cars as company cars for the ASMs. My method was to tell them that the ASM with the best month's sales performance could have the first pick of the cars in the showroom, the second-best could choose next, and so on until the remaining car would be left for the ASM with the poorest sales performance. This would enable me to change the display each month without almost any personal effort.

The first time I did this, Alan Tweed had the poorest sales result and the car not chosen by any of the other ASMs was an Austin Allegro Equipe. This was a limited-edition two-door special model with overoptimistic marketing as being the Allegro that had 'Vroom!' All Equipes had a silver body colour with 'go faster' orange and red stripes down each side to reflect the car's 1750-cc engine, unique alloy wheels and five-speed gearbox. I don't how many Allegro Equipes have survived but assuming there are some they are likely to be the rarest of all Austin Allegro models produced.

Needless to say, it wasn't the sort of car the ASMs wanted to be seen in, but Alan had no choice. The following month the cars were chosen by the ASMs once again in order of the ASMs' sales performance, and the cars included another Allegro Equipe. You can probably guess what happened. Alan Tweed was left again with no option but to take another

Allegro Equipe, owing to his area's poorest sales performance. I can remember him asking me if there was any way he could have another car as he claimed the Equipe was damaging his credibility with his distributors and dealers, and that it would get worse if they found out he was having to drive another one. The rules were the rules so he had to take a second Allegro Equipe as his business car.

In March 1980 the Zone 5 Regional Office was relocated from Piccadilly to new offices in Ealing in West London. I had been made aware of a possible office move in September the previous year. In practice I now had a shorter drive from home to Ealing than Piccadilly so that was good news.

The year 1980 was notable for what I consider to be one of the most bizarre car launches I attended in my long career in the car industry. The car being launched on 12th May to the distributor and dealer network was the Morris Ital. The car's name came from Giugiaro's Ital Design studio in Italy, which had been employed by BL to redesign the Morris Marina, a car which had been produced since 1971 and which by now should have been replaced by a new model. However for financial reasons we got a major facelift. The Marina name was dropped on the orders of Michael Edwardes and the Ital was launched to the UK public on 1 July 1980. It had revised exterior styling, noticeably at the front and rear, but retained the Marina's 1.3- and 1.7-litre petrol engines. The dashboard and interior of the Marina were also carried over largely unchanged. In my opinion the Morris Ital still looked very much like a Morris Marina.

Dealer group heads, distributors and dealers were invited to attend the Ital launch on 12th May. It stated they should arrive by 9 a.m. to check in at the Penta Hotel at Heathrow Airport. From there they would be transported by coach to the Cricklewood Film Production Studios in North London where the car launch would take place starting at 11 a.m. Following this and lunch they would be transported back to the Penta Hotel for a test drive of the Ital, then dinner with cabaret and an overnight stay at the hotel.

Not surprisingly I received a number of requests from dealer group heads who wanted to go direct to Cricklewood, because it was more convenient for them to do so. However, they were all told they had to go to the London Airport Hotel first. This generated a number of strong complaints especially from those group heads who had their London offices much closer to Cricklewood than Heathrow.

I distinctly remember it was a very hot day. The journey, which was around 17 miles along the M4 motorway and then the A406 North Circular Road, should have taken around forty minutes. However, there were major road works on the North Circular Road that day which created serious traffic jams along the road. The result was that all the coaches arrived at Cricklewood Film Studios very late. That wasn't the end of the problems.

We were all ushered into a very large wooden hut which had been constructed to convey the idea we were going to get an army-type briefing. Without any air conditioning the interior of this large hut was unbelievably hot and humid. The car launch delivered by our top management was indeed on a military theme with the enemy being the competitor manufacturers of Ford and Vauxhall. In the film that was shown the enemy cars were chased around an airfield with gunfire sounds. At the end of this attempt to mirror an air battle the Ital came to a stop, the driver got out and stuck a Ford logo on the driver's door, alongside other Ford logos, to imply another 'kill'. The proceedings concluded with a very late lunch of sausage and mash to the somewhat bemused and far-from-happy audience. I know there was a shortage of food because I and many others from the company didn't get any. I went with a few other employees to a nearby pub for lunch.

After the launch we had to endure a very long and delayed return journey along the North Circular Road which put us even further behind the time schedule. So by the time we arrived back at the Penta Hotel for the test drive we met the evening rush hour, and Heathrow Airport at rush hour was not a good place to run a test-drive event.

The Ital was unable to arrest the gradual slide to a lower Austin-Morris market share in the four years until 1984 when it was produced and sold in the UK. It was very much outdated by the time it was launched, and with build-quality problems it soon gained a bad reputation. In July 2008, it ended up second in a poll of 'The Worst British Car Ever' conducted by *The Sun* newspaper, narrowly behind the Austin Allegro!

From the time I came to the regional office in London and into 1980, I became increasingly involved in the monthly wholesale process and it was normal for me to meet with the directors of dealer groups such as Maurice Bayliss at Henlys and Alan Caffyn at the Caffyns Head Office in Eastbourne. It continued to be a struggle to get the distributors to order vehicles in the numbers and model mix that we required to meet the production schedules for the total BL model range.

Mentioning Maurice Bayliss reminds me of one lunch with him and his directors in Eton. Lunch with Maurice and his colleagues was always enjoyable. Including Zone 5 senior managers, there were about ten of us in total. Tony Coleman, a Regional Director of Henlys, was standing up and speaking on some industry topic and while gesturing with one hand knocked a full bottle of red wine over in my direction, where I was seated opposite him. Almost the whole bottle of red wine went over me, my shirt and my suit. Before I could say anything Tony picked up the full jug of water in front of him and threw the whole jug of water over me. He then said, 'You will hate me now but you will be pleased I did it later' I was speechless and in fact there was nothing I could say. Lunch on this particular day then continued almost as if nothing had happened. It was difficult to explain to my wife Annette about my very wine-stained appearance when I arrived home but that was the honest explanation I gave her.

In early 1980 I booked a family holiday with a difference. Later, on 2nd June, we travelled on British Airways to Kenya for our most adventurous holiday to date. It was our first holiday outside Europe and Kenya would not have been most family's holiday destination at that time. In addition, Gilly was only six years old. Holidays to Kenya were not widely available or affordable to most families and those that did consider Africa at that time would not have ventured beyond Tunisia.

We flew to Mombasa. On landing the following morning we travelled 20 miles north to the Sun'n'Sand Hotel on Kikambala Beach on the east coast, where we would be staying on a full-board basis for two weeks. All the people staying there were British and had come on the same flight as us. A day after we arrived the British Airways travel representative came and told us we could not possibly come to Kenya without going on a safari. A safari was not part of our holiday package and after a while he convinced us and the Harris family to take a short safari. However, I did not have enough money on me to pay for it and neither did Peter Harris and so the two of us went into Mombasa to get some local currency.

Very early on Sunday, 8th June the Sparrow family of three and the Harris family of four set off in a VW Microbus people carrier and travelled inland maybe 150 miles to the Tsavo West National Park. Once there we took photographs of wildlife, which we were able to do through the open roof of the VW. We continued our travel west into the Amboseli National Park, one of the best wildlife viewing experiences anywhere, which provides spectacular views of Mount Kilimanjaro, the highest single free-standing mountain in the world. It became an objective to take photos of the animals with Mount Kilimanjaro in the background. One of my favourite photographs of Annette, Gilly and me is the one in Amboseli with Mount Kilimanjaro standing tall behind us.

It had been a wonderful experience and I am very pleased that we were persuaded to go on this short safari. I did wonder afterwards what would have happened if our VW Kombi had broken down because for most of the time on our safari we were in a single vehicle without any means of communication in areas where we did not see any other vehicles.

Back in the UK we had our annual Regional Review with Peter Johnson, our Sales Director. The sales performance levels in the Zone 5 region were never good enough in the eyes of top management, even though the Regional Office team and the distributor and dealer network were working as effectively and as hard as they could to achieve the sales objectives.

September 1980 was memorable because the company would be launching a brand-new car. This was a car upon which Michael Edwardes and his top management said the company's future depended, as the company continued to have serious financial problems and there were fears that it could go out of business. The car hailed as BL's saviour was the Austin Metro, although it was launched to the public on 8th October as the Austin mini Metro.

Even before the launch there was enormous public interest in the car as details appeared in the press. In its wisdom the company's marketing department decided to carry out the launch presentations on board a cruise ship, the MS *Vistafjord*. The plan was for everyone to board the ship in Liverpool and for the ride-and-drive to take place on the Isle of Man. Five rotations of this event were arranged to cater for the large number of invited attendees.

I arrived in Liverpool on 10th September to learn that owing to very rough weather in the Irish Sea, the previous rotation had failed to get people ashore onto the Isle of Man. The omens were also not good for us. We embarked on the *Vistafjord* on the afternoon of 11th and we all attended the first conference which covered our plans for other models in 1981. According to the schedule, next day we should have left the ship on smaller boats for the ride-and-drive – everyone's first experience of driving the Metro.

However we were told by the pilot that there was absolutely no chance we could do this owing to the very rough sea and gale-force winds. One alternative was to watch the disaster comedy movie *Airplane!* in the ship's cinema. The word disaster seemed appropriate! This was another clever launch plan on paper, which like the Morris Ital failed in its delivery for the many who were unable to drive the car as planned.

A major TV advertising campaign was created for the Metro with the headline 'a British car to beat the world'. The advert also featured similar-sized competitor cars such as the Fiat 127, Renault 5, Volkswagen Polo and Datsun Cherry as 'foreign invaders' and the TV advert spoke of the Metro's ability to 'send the foreigners back where they came from'. Those old enough will remember the TV advert where a number of Metros are parked on the edge of the cliffs at Beach Head, near Eastbourne on the south-east coast of England, facing container ships out at sea laden with foreign cars that would go back out to sea.

The Metro was launched to the public on 8th October 1980 and all distributors and dealers held their own individual launch events. I supported Tony Ball, one of our Main Board Directors, at two dealer launch events in Farnham, near where he lived. Also there was Tony's son Michael, now a well-known singer and actor. The Metro was popular with car buyers, and during its first few years it was the best-selling small car in the UK, before being overtaken by the Ford Fiesta in 1984.

At the end of 1980 I received an interesting phone call from Head Office. Lady Diana Spencer was being romantically linked in the press with Prince Charles, Prince of Wales. The call I received required me to immediately produce, to an exact specification, a dark-red Austin Metro for Lady Diana. I made an immediate phone call to the manager responsible for vehicle supply at Longbridge. A dark-red Metro was identified on the assembly line and very shortly after being produced and inspected was on its way to Thomas Day Motors of Fleet, Hampshire.

On 24th February 1981 the engagement of Lady Diana Spencer to Prince Charles was announced. A month later on 24th March my mother and I accompanied my father to Buckingham Palace where he received the MBE from the Queen for his services to local government. He had served on the Wheathampstead parish council and the St Albans rural district council since the 1950s. As we entered the inner area of

Buckingham Palace I saw a dark-red Metro parked there. It was a car that probably took one of the shortest ever periods of time from its order to its delivery in meeting a customer's urgent requirement.

On 26 December 1979 Michael Edwardes had signed a collaboration agreement between BL and Honda that was aimed at securing a future for the British company. Based on the Honda Ballade, the new car was named the Triumph Acclaim and was launched to the public on 7th October 1981. The end of Triumph Dolomite and TR7 sportscar production meant the Acclaim was the only Triumph-named car after 1981, and it was launched as a car 'totally equipped to triumph'! The Acclaim represented some badly needed improvement in the image of BL, as the car had good reliability and build quality from the start. The Acclaim was the start of a range of Honda-based, Rover-badged cars which BL, Austin Rover and Rover Group would produce in the 1980s and 1990s. For a while in 1982 and 1983, the Acclaim featured in the top-ten-selling cars in Britain, the first Triumph car to achieve this since records began in 1965. At the end of the year I met with management at Land Rover to discuss a job opportunity in what was now a separate company within BL. This would start and lead to one of the most enjoyable and successful periods of my career in the car industry.

Chapter 8

# Land Rover Ltd 1982–1984

**ALTHOUGH THE LAND ROVER BRAND** originates from the original 1948 model, Land Rover as a company was only formed 1978. Prior to this, it was a product line of the Rover Company which was subsequently absorbed into the Rover Triumph division of BL. Land Rover sales responsibility continued within BL until 1979–1980 when a new dedicated Land Rover sales and marketing function was established to develop the brand.

In 1982 I joined Land Rover Ltd initially in Franchise Development, a role based at the Land Rover factory in Solihull, south-east of Birmingham. For the first time my job would include Europe as well as the UK. Commuting from Camberley to Solihull in 1982 was not pleasant or quick. The journey was 120 miles and it took around three hours.

A revised UK franchise plan was being developed by Land Rover management and initially I was involved in reviewing some UK locations where it was considered some additional franchise coverage would be useful to grow sales. These were primarily in rural areas and I recall there was a view that recruiting some dealers with agricultural equipment/tractor franchises might increase Land Rover sales as they had direct access to farmers who were potential Land Rover customers. At that time most Land Rover franchises were still linked with BL Car franchises in city and urban locations.

I had my first experience of driving a Land Rover vehicle at Land Rover's off-road test track at Eastnor Castle, near Ledbury, in Herefordshire. Along with some other managers I was given some off-road driving instruction before driving off in a 109 V8 station wagon on the black route – one of the most demanding routes there.

Driving for the first time in the right low-range gear down a steep slope in a Land Rover fully laden with other humans and without power steering was an interesting experience. However, I learned a lot that day about the go-anywhere capability of the traditional Land Rover. It would go anywhere, providing the driver was competent enough to drive it anywhere, even if you were up to your neck in water, although I never had to test myself in that extreme situation.

In November we sold our house in Camberley and began house-hunting for the fifth time. As I was based in Solihull we concentrated our search within half an hour of my office. We soon identified a house in Knowle, just south of Solihull and ten minutes away from the Land Rover factory. It was not an ideal property and it would be the first we owned that was not new. We moved into 1 Trehern Close, Knowle on 19th February 1983.

Shortly after that I was given a job that would become very important in progressing my career in the motor industry. I was delighted to be told by John Cunnane, Director UK Sales Operations and UK and Export Service, that I would be offered the Manager UK Sales Operations position, confirmed by Jack Reardan, the main board Sales and Marketing Director.

John Cunnane was a great boss to have. I remember some of his first words to me. He said very clearly: 'I will direct you but you will manage the department.' It was quite a simple business statement, based on our respective job titles, of how our working relationship would be, and this was how things worked successfully between us. Six weeks later in June 1983 I also took responsibility for the operational implementation of Land Rover franchising strategy, also reporting to John.

For me working with John was a very enjoyable and successful period during the rest of the 1980s as we helped develop and grow Land Rover sales in the UK. This was an important period of stability and continuity for both of us because in addition to our sales responsibilities we also spent much of our time together working to develop a more exclusive and dedicated UK dealer network that would be the foundation for the future success of the Land Rover franchise and the company's business.

Within days of starting my new job I attended my first Land Rover wholesale meeting. What a difference this was to the monthly wholesale process I experienced in my last BL Cars sales job in London. In BL Cars there was a continual battle each month to get the distributors to order the very large numbers of cars we needed to satisfy the manufacturing and production requirements, because they already had enough cars in stock and on order to meet their expected sales, without ordering any more. However, in Land Rover in 1983 the situation was very different. There was no problem getting the distributors to order the relatively small numbers of traditional Land Rovers and Range Rovers on offer.

The annual UK sales target for 1983 was 5,800 short- and long-wheel-base Land Rovers and 2500 Range Rovers, so on average the total monthly wholesale order was barely 700 vehicles and some of that total had to be made available to satisfy confirmed fleet customer requirements. In 1983 there were 350 Land Rover franchise holders in the UK, most of whom operated the Land Rover business alongside their other BL Cars franchises of Austin Morris and Rover Triumph. With an average of around twenty-four new Land Rover/Range Rover sales for each dealer a year the Land Rover business at that time was not significant for most dealers and for many it was treated merely as a nice profitable 'add-on' to their main volume car activities.

The UK Sales Operations team I inherited when I started my new job was as follows. There were five Regional Sales Managers, namely Charlie Sneddon, Martin Luxton, John Evans, Peter Cleaver and Bob Hester. The Sales Administration team comprised David Carpenter, Manager, supported by Jim Cooke and Will Hope. My secretary was Audrey Leaver. It was a small but experienced team.

The Regional Sales Managers were effectively home-based while the rest of us worked from offices adjacent to the vehicle assembly line. My office was very small and there was a hatch in the wall which enabled me to speak to Audrey on her side of an adjacent office, or for her to speak to me. In practice this was an antiquated way of communicating but it seemed acceptable then.

One of the first things I did was to collect my new company car which was a Range Rover, registration number FAC156Y, in a gold body colour. It was a four-door model that had been launched in 1981 and it had an automatic gearbox using the three-speed Chrysler 'Torqueflite' which had been introduced a year earlier in 1982. The Range Rover I drove in 1983 was a long way from being today's luxury vehicle of choice for the rich and famous, including royalty, actors and footballers.

In May 1983 I had my first meeting with Tony Gilroy, Land Rover's new Managing Director, to discuss a Range Rover campaign. I don't remember the meeting but I do recall being told about the new Managing Director I was to meet for the first time and his so-called fearsome reputation.

The following month I was to witness the personality and leadership style of Tony Gilroy while attending the monthly Sales Operating Review – known for short as SOR – meeting in the boardroom. This meeting was chaired by Tony and attended by all the Directors responsible for or with an interest in Land Rover's global sales.

At this SOR I attended as a deputy for John Cunnane and I sat directly opposite Tony. Everyone else had deliberately avoided sitting there opposite him! At the start of the meeting he asked me why I was there and I said I was deputising for John Cunnane. He said, 'We don't have deputies here.' I was about to get up and go but he immediately just said to stay!

The meeting was an eye-opener. Tony asked each Director for a report on current and future sales prospects and I quickly learned that what he wanted were facts not opinions. As an example at this SOR meeting Tony asked a Director about the likelihood of Land Rover obtaining a contract to supply Land Rovers to an African country's army. At the time there was some doubt about the continuity of that country's government so the response to Tony was that the situation was unclear. There was no better response when the Director was asked by Tony whether or not success in supplying was probable. At a later SOR the same topic arose and when there was a similar answer to the MD's question. Tony gave a definitive answer, 'Because I have taken the trouble to find out.'

When it came to my turn I gave a factual statement on the current UK sales performance and said that we would achieve our sales objectives. I had addressed the MD as Mr Gilroy but was immediately told very strongly to 'call me Tony'. I attended other SOR meetings in John Cunnane's absence and made sure I had fully prepared and memorized what I was going to say. A few Directors attended the SORs with piles of folders and documents but I knew it was very unlikely they were going to be given the chance to refer to them. It seemed to me this was displaying a degree of weakness and would give the impression they were not fully prepared. I just wanted to look straight at Tony and give him the factual statement I knew he wanted, even if this was not all good news.

The sales achieved in the UK made a very significant contribution to the company's financial performance and so I quite often received a phone call from Tony out of the blue with him asking some sales-related question. I did not always have the answer, but I knew Tony wanted facts so on these occasions I said, 'I will call you back shortly' – or however long it would take me to get a factual answer. I was always as up-to-date as I could be with all the relevant sales performance figures, forecasts and stock levels. However, knowing I could receive phone calls from Tony at any time meant I took great care to try and anticipate every sort of question he might ask, especially those related to achieving month-end sales targets, state of the market and so on.

I can say that I greatly appreciated this direct and clear communication from Tony and I can say quite definitely that he made me a better manager, because through him I better understood the importance of my role and the positive impact my UK sales team and I could make on the company's success.

Land Rover at that time through Tony's leadership started to introduce a programme of new vehicles that would be necessary to grow sales. In 1983 the Land Rover 110 was launched, replacing the previous 109-inch model. The new vehicle had the coil-spring suspension of the Range Rover in a new, stronger ladder-chassis frame. Other features included a five-speed gearbox, front disc brakes, a one-piece windscreen and optional power steering. The extended-wheelbase Land Rover 127 also appeared in Crew Cab form. The same year the Range Rover was upgraded with a five-speed manual gearbox and other improvements. Annual production was around 12,000 vehicles.

In the early 1980s Land Rover vehicles were the biggest-selling Land Rover models in the UK and customers in the agricultural sector were the biggest purchasers of the product. Accordingly, having a presence at the UK's major agricultural shows was an important means of marketing the vehicles. Land Rover had a sizeable company stand at the major national shows displaying and promoting the full range of its products. Regional shows would be supported, though there was a greater commitment here for the local dealers to operate and man a stand.

In June 1983 I drove to Scotland to attend my first Royal Highland Show at Ingliston, to the west of Edinburgh. This was by far the most important agricultural show in Scotland, but not in the Highlands, despite its name. I remember my first drive to Scotland very well because my occasionally 'gutless' Range Rover struggled to climb any steep gradients on the M6 motorway, and slowed down dramatically in the process, especially through the uphill parts of the Lake District. Although I started with a full tank of petrol my fuel gauge was almost on empty and the vehicle had averaged just under 13 miles to the gallon by the time I had travelled the 218 miles to Gretna Green. The whole 330-mile journey to the hotel took five hours.

The Land Rover staff manning the Land Rover stand at the Show would be based at the Inchyra Grange Hotel, at Polmont, near Falkirk. The hotel would also be the venue for the annual Scottish Land Rover dealers' luncheon. For me this turned out to be a memorable event during my few days there. The previous year Land Rover had hosted a dinner at the hotel for the dealers and I was told there had been an abnormally high level of alcohol consumption even by Scottish dealer standards. Apparently one dealer had become very inebriated even before dinner had started and during the first course his head dropped into his bowl of soup. The story goes that a dealer sitting next to him had the presence of mind to lift his head up, thereby preventing the dealer from drowning in his soup. I wasn't there to witness this improbable event but I was told it really happened.

After what occurred at this dinner the previous year it was decided that a luncheon might be more appropriate but things didn't turn out that way. Prior to lunch drinking started at the bar and during lunch alcoholic drinks were provided. At the end of lunch I stood up to make a short speech on Land Rover's sales performance in the year to date and thanked the Scottish dealers for their achievements and support. However I noticed the dealers were gradually taking less interest in what I was saying. This was possibly owing to the fact I saw that the waiters were putting bottles of scotch whisky on the tables and that in some instances the bottles were being immediately put under the table and waiters were being asked to provide more bottles to fill an empty spot on the table.

Fortunately my speech was a fairly short one and so after I had finished the drinking could continue uninterrupted. Many dealers were staying overnight with plans to visit the Royal Highland Show the following day so it was no surprise that for them drinking continued. Later that afternoon one dealer was found asleep, and very drunk, outside in one of the hotel's flower beds. This luncheon was my first but not my last experience of some of the Scottish dealers' incredible drinking capacity.

Summertime meant there were also many important regional agricultural shows. I was invited to most of these to show my support but because I had to manage my work priorities, if I attended any it was because I could include visits to local Land Rover dealers in the same itinerary. I was keen to visit as many dealers as possible in my first year not only to motivate and encourage the dealer teams by face-to-face communication, but also to judge the dealers' ability and commitment to meet our franchise plans and business requirements.

During the summer of 1983 John Cunnane and I started to hold business review meetings with the major national dealer groups in order to outline our policy and plans for a more dedicated and exclusive dealer network in the UK, to discuss their sales performance and determine their level of commitment to developing the Land Rover franchise and their potential investment intentions.

These dealer groups tended to operate Land Rover within premises that were shared with and were more geared to selling and servicing the volume franchises of the overall BL Group, such as Austin, Morris and Rover cars.

It would take time for most of the major groups to commit investment to creating exclusive Land Rover dealerships, especially until Land Rover was able to demonstrate this would be a profitable long-term opportunity. Jaguar Cars had introduced a policy aimed at creating an exclusive dealer network wherever possible and to achieve this had started the process of requiring its dealers to invest in exclusive premises. Dealers and dealer groups naturally decided investing for Jaguar was a more logical business route to go at that time rather than in the much smaller opportunity Land Rover offered.

In September I flew to Edinburgh to spend four days visiting most of the dealers north of the city. Charlie Sneddon picked me up from the airport and on the first day we visited Heron in Perth, followed by

Macrae & Dick in Inverness, Nairn and Elgin. That evening we stayed at the Rothes Glen Hotel, south of Elgin, so we were conveniently located for the dealer visits the next day.

The first visit the next morning was to Frank Ogg at Aberlour on Spey. I remember our meeting well. On arriving we met Frank the owner of the business at his home near the dealership. He said he would get us a drink and came back into the room not with coffees or tea but with a jug of whisky which he poured into some glasses. He introduced this to Charlie and me as very-high-proof malt whisky you cannot buy in the shops. I drank some and immediately knew I had never had alcohol as strong in my life.

Having consumed some of Frank's whisky I was very grateful that Charlie was doing the driving that day. The next day I was back to normal and visited dealers in Stonehaven, Forfar and Dundee. The trip concluded with visits to the Carmichael Group's dealerships in Stirling, Falkirk and Edinburgh. However it was the meeting with Frank and his whisky that I will always remember.

In Land Rover we realized that we depended on too many small dealers, very few of whom specialized in the 4x4 business and most of whom had other car franchise priorities. This was not going to be the best way to grow a specialist vehicle business in future and so we carried out a detailed franchising review of the UK to determine the optimum number of dealers to handle the future sales volumes of new vehicles and the associated service and parts business. Land Rover Parts had not long before been set up as a separate company and so our franchising review included meetings with Alan Simpson, the Managing Director, and his team.

At the end of our review the following franchise strategy was determined. The UK would establish a network of exclusive and dedicated dealers wherever possible, operating as main dealers with a direct supply of vehicles and parts. Each main dealer would have a geographical area of responsibility with sufficient sales and after-sales potential to justify investment in an exclusive franchise operation and to give a good return on that investment. Indirect retail dealers would only be retained largely to provide after-sales coverage for customers in very rural areas where a main dealer investment could not be justified.

One of our first steps was to terminate a number of very small retail dealers that were very clearly superfluous to requirements and whose Land Rover business could be picked up by larger Land Rover dealers in the same area. These actions were carried out on a lowkey case-by-case basis as there was no need to destabilize the dealer network with any announcements of a major refranchising initiative at that time.

Even this relatively lowkey activity caused concerns within the Austin-Rover part of BL as we were invariably terminating a dealer from within the overall company. And quite often it was a dealer that had already been terminated as part of Jaguar Cars franchising policy. So I attended a regular Franchise Operations meeting so that we could discuss franchising issues and in particular Land Rover's franchising intentions. Over the next few years the latter would lead to some internal politics as Austin-Rover's top management felt Land Rover's increasing actions to seek dealer exclusivity were undermining their dealers, especially in terms of profitability and viability.

Some fairly basic 'operating standards' had been introduced and 'recommended' for all Land Rover franchise holders in the UK. There were just fourteen standards in total. For example, 'Every outlet must identify and train one salesman responsible for 4x4 sales.' However, in practice this person could spend very little time selling Land Rover vehicles. There was a similar operating standard for service, with, for example, 'Every outlet must identify and train one productive.'

However, the process of developing true specialization in the Land Rover network only started with the appointment of the first 'Land Rover Centres'. These were existing main dealers that already had a high level of specialization in the Land Rover franchise.

Towards the end of 1983 we had some more internal meetings to discuss the sort of minimum franchise standards we would require specialist Land Rover dealers to meet to go alongside our franchise strategy. One of those standards we discussed at length was a requirement for all our dealers to have a number of demonstrators to enable potential customers to test-drive vehicles across the model range. We knew that customers were more likely to buy having experienced and driven the product first, so we had to put in place a demonstrator programme that made this a normal part of the dealer sales process, which meant providing some financial support.

In November 1983 management was briefed by Tony Gilroy on the opening of the Rover SD1 plant so that in future all Land Rover manufacturing-related activity would take place at the Solihull factory. Much of the factory was in need of investment and redevelopment and there was still a semi-cottage-industry feel about the way the vehicles were assembled. In addition, the main Directors' office block still had camouflaging on the brickwork from the Second World War.

The month of December meant a few days attending the Smithfield Show at Earl's Court in London. This was another large agricultural show and I recall we had to get special permission from the show organizers to display some of our vehicles as they were not 'agricultural equipment'. Most of the Land Rover dealers who had agricultural equipment franchises attended and this gave me opportunities to have short business meetings with them in our office within the main stand. It was a little like holding a doctor's surgery.

I remember conversations I had on the stand with Land Rover customers. Most had come on the stand to make some point or complaint about the poor quality or problems with their vehicle. It could be irritating issues such as squeaks or rattles, or the less-than-perfect fit and finish of the trim within the vehicle. However, having acknowledged that on some of these points it was down to the inherent design of the vehicle, the customers invariably said they would still buy another Land Rover. The loyalty of the customer was a critical factor in the success of the Land Rover business. Most customers who experienced problems were prepared to tolerate and put up with product quality that was less than that offered by the Japanese competitors' vehicles that were coming onto the market.

However, as we gradually developed the Range Rover into a more luxury specification, which would appeal to luxury car customers, we would be faced with having to offer luxury car services to the customer of the sort offered by Jaguar and Mercedes, and this included offering courtesy cars when the customer's vehicle was off the road or in for routine servicing.

I realized that an important part of my job was to be seen by the dealer network, so for at least two days of the week I planned to be out with my Regional Managers visiting dealers, discussing sales matters of mutual interest. It was also an important way of having a 'good feel' for how things were going as I insisted on getting honest answers to questions on the market, pricing, product, competition, future prospects and so on.

At the start of 1984 John Cunnane and I visited most of the major dealer groups to carry out a business review that included sales performance of each outlet and franchising matters. These meetings once again included Dutton-Forshaw Group with Andrew Love, Henlys Group with Ben Doole and Pat Shewring, Kennings with Peter Dean, Lex Group, where we met John Tinker and Richard Martin, Mann Egerton Group with Brian Bleaney and Archie Clayton and Wadham Stringer, where our meetings were with Les Taylor, the Managing Director.

In May 1984 I attended the Land Rover dealer council meeting at Solihull. This was a joint Land Rover Ltd and dealer network structure aimed at discussing the key policy, strategy and operational issues of the time. There were two dealer council members for each of the five regions, so ten in total. The future franchising policy, franchise strategy and franchise standards became main subjects at these meetings because the move to a more exclusive dealer network would involve a relatively high degree of rationalization and high levels of investment for the fewer long-term outlets. Where these important actions were going to be implemented, these could hopefully be explained, discussed and agreed with the dealer council members on behalf of the whole network before communicating them to the dealer body.

One matter of mutual interest that emerged during this period was the impact of non-franchised Land Rover specialist dealers. Two in particular that were located in the South-East of England were obtaining new vehicles and having an adverse impact on the franchised dealers' business. These two companies were Townley Cross Country Vehicles, based in Eltham in South-East London, and Hunt Grange Vehicles based at Lamberhurst in Kent. Townley Cross Country Vehicles, also carried out a variety of Range Rover conversions especially for very rich customers in the Middle East.

These two companies specialized exclusively in selling and servicing Land Rovers and this was their appeal. They were able to source new unregistered Land Rover vehicles at attractive prices from official Land Rover dealers, especially from France and Belgium. It was no coincidence that one of the biggest Land Rover outlets was in Boulogne on the north coast of France. As exclusive 4x4 operators and with businesses dedicated to Land Rover vehicles these two companies provided very strong competition to many official dealers in their areas who were not exclusive Land Rover dealers. In the end we decided these two exclusive Land Rover specialist dealers would be better working with us and their customers officially and so they were appointed into the franchised dealer network.

The aforementioned dealer council meeting of May 1984 included discussions on new models, because the following week the new Land Rover 90 and Range Rover Phase II were launched to the UK dealers over a period of four days at Eastnor Castle, near Ledbury in Herefordshire. The Land Rover 90 was introduced replacing the previous Land Rover 88-inch model. The 90 had new doors with wind-up windows, a full-length bonnet, revised grille, plus the fitting of wheel-arch extensions to cover wider-track axles. Other changes included coil springs, offering a more comfortable ride, and improved axle articulation, a permanent four-wheel-drive system derived from the Range Rover, featuring a two-speed transfer gearbox with a lockable centre differential, a more modern interior and a taller one-piece windscreen. Following a successful limited edition, a more highly specified Range Rover Vogue was introduced at the top of the model range.

At the end of the second day's launch I drove home from Eastnor Castle to pick up my wife Annette. Having done so we travelled north up the M6 motorway to Stockport where Hollingdrake, part of the Heron Motor Group, was having a new showroom opening for its Jaguar and Land Rover franchises. I was met by John Turner, Managing Director of the Heron Motor Group, who told me that Roger Putnam, Sales and Marketing Director of Jaguar Cars, would be speaking on behalf of both the motor manufacturers. That was fine for me. However he then surprised me by telling me that I would follow Roger by compering the fashion show. I was not prepared for that but John just said I should speak to the lady running the fashion show and she would give me some advice and some words to say.

I found out that Margaret of Wilmslow (I think that was the name of the company) was providing the models and the fashion wear. She gave me some words which I wrote down on a piece of paper. After Roger Putnam had spoken it was my turn. I introduced the fashion show to the large number of people in the showroom that also included Gerald Ronson, Chairman of the Heron Motor Group, and his wife Gail by saying the models were wearing everyday dresses at affordable prices from Margaret of Wilmslow. Even in 1984 each garment cost several hundred pounds and I thought what I had to say, especially about affordable prices, was somewhat exaggerated. However, the people in the showroom seemed pleased with the fashion show and I had done a reasonable job at promoting some fashion. This would be the first of many new Land Rover showroom openings I would attend over the following years and fortunately I was never put in the same situation.

In July 1984 I was involved for the first time in the company's graduate recruitment process and spent some time interviewing potential employees. At the start of September one graduate trainee joined my department for six months and I could tell I had been given a very talented young man with great potential. His name was Richard Baker but unfortunately for us he didn't stay long at Land Rover. He moved to Mars then ASDA and nineteen years after his short time at Land Rover he had risen to become Chief Executive Officer of Boots.

At the October Dealer Council meeting we outlined for the first time our plans to introduce a franchise development programme called 'Business Builder' into the UK dealer network. The programme would operate with Land Rover Ltd retaining 3% of the dealer margin on all Land Rover 90 and 110 models, leaving the dealers with 13% net discount, and retaining 3.5% on all Range Rovers, leaving a 15% net discount for the dealers. The total of all these retained margins would be held by Land Rover Ltd in a fund and then paid back periodically to the whole dealer network in franchise development programmes and initiatives, such as corporate identity, that would raise dealer standards and benefit the dealers' business.

We had learned the business benefits of such a programme in raising dealer standards from Jaguar Cars. We regarded Jaguar as a benchmark for a specialist car franchise and although we were not producing or selling luxury cars, we were aiming to develop a strong and profitable franchise that all the best dealers and dealer groups would want. It would not be until January 1987 before the Business Builder programme was finally launched.

Also by 1987 we created marketing areas for the Land Rover franchise so that each franchise holder had a defined area of responsibility with sufficient business potential to justify an exclusive dealer operation. In addition the franchise standards would be reviewed and refined to reflect dealers that were 'centres' or main dealers, operating on a direct account with Land Rover, and those that were retail dealers, obtaining vehicle supply via a main dealer. We had an increase in our sales objectives for 1985. The figures were 7,100 Land Rovers and 3,200 Range Rovers, so we had for the first time a target of over 10,000 sales. This was almost 2,000 more than the target in 1983, but we entered 1985 with 270 dealers compared to 350 two years earlier. With an improving product range the sales task my sales team and the dealer network faced was realistic and achievable.

## Chapter 9
# Land Rover Ltd 1985–1988

**IN EARLY 1985 I STARTED** to become involved in Land Rover Ltd's product development. This meant I was invited to product viewings at Drayton Road in Solihull where vehicles at various stages of evolution and development were subject to management scrutiny and evaluation. I thought it was good for me to be able to comment on these vehicles since my sales team and I would in due course be responsible for selling them.

In March and April 1985, I was involved in escorting a number of dealer groups including their key staff and special customers on a series of Solihull factory tours. Land Rover was at a stage where this could be a positive initiative in strengthening its relationships with its dealers and key customers. Seeing the various stages in the production of a Land Rover or Range Rover had a very beneficial impact on all these visitors. I remember we also included a couple of factory tours for the police.

On 1st May 1985 I gained responsibility for UK fleet sales and for Land Rover sales in Ireland. For several years there had been operational sales problems resulting from the fact that pricing variations between Land Rover vehicles in Ireland and Northern Ireland led to many vehicles crossing the border, with many being handled by the non-franchise trade. As both markets took the same right-hand-drive vehicles, with similar specifications, Tony Gilroy decided that now was the time for some sales and marketing harmonization in these two markets. It was now possible as well as desirable.

Reg England, UK Service Manager, and I were asked to make a short factfinding visit to Ireland and so we went to Dublin. There we were met by Dene McQuaid, who was responsible for Land Rover sales within the Austin-Rover National Sales Company. We visited Stuarts of Dublin, one of the larger Land Rover dealers, and in the afternoon had an operational review of the Land Rover business with senior Austin-Rover Ireland management.

The following day Dene took us to visit Hassetts Motor Works, a Land Rover dealer in Thurles, a town in County Tipperary. I remember the visit because on arriving Dene expressed surprise that the owner was wearing a suit especially for our visit. Apparently, Mr Hassett never wore a suit. I also recall we had some difficulty getting into the parts department because the parts manager would not allow anyone into his domain. Once he had relented under orders from the owner and we entered the parts area, I was surprised to see that many of the smallest parts were stacked on shelves in rows of labelled jam jars. My planned questions about parts stock control systems became superfluous! As a result of the visit we decided to develop a small, more dedicated Land Rover dealer network based on UK franchise standards.

In early August of 1985 we made our first visit as a Sparrow family to Gatcombe Park for the horse trials. Land Rover sponsored the Range Rover team, and in due course this brought me into contact with Captain Mark Phillips. After establishing a relationship with him in 1980, Land Rover funded a programme of training sessions and bursaries for young riders, which he created and ran, and which was branded 'the Range Rover Team'. Two years later Land Rover became one of the original sponsors of the British Open Championship which first ran at Gatcombe Park in 1983.

In August 1985 I travelled with Annette and our daughter Gilly to Scotland once more. This time the Land Rover marketing department had agreed to sponsor some events at the Lonach Highland Games, which are held annually at Strathdon, some 45 miles west of Aberdeen, and we would present prizes to the winners of the Land Rover events.

The Lonach Games are a traditional Highland gathering with events such as tossing the caber, hammer throwing and so on, and also include a competition of Highland dancing. The Lonach Highlanders are members of the Lonach Highland and Friendly Society, made up of men from the Strathdon area of Aberdeenshire. The principal aims of the Society are the preservation of Highland dress and the Gaelic language, to support loyal, peaceable manly conduct and the promotion of social and benevolent feelings among the inhabitants of this district. Led by their patron of the Forbes clan, the Lonach March features the men of Lonach in full highland dress and with banners flying. The Lonach March takes place in the morning before the games start and at various points along the route from Bellabeg to Strathdon the column stops for a 'wee dram'.

Once the march is completed the men of Lonach have lunch together in the Lonach Hall. I was told by Gibbie McIntosh, Secretary of the Society at the time, that I would be granted the rare privilege of joining them for lunch, something not normally accorded to an Englishman. In fact as an obvious outsider wearing trousers and a sports jacket I felt I was an invader of the Highlanders' privacy and out of place in a building where everyone else was in full highland dress. However, I enjoyed the experience of being with the Highlanders during their special annual lunch and of hearing the after-lunch address from their patron Sir Hamish Forbes.

During the Saturday morning we had been invited into the organizers' tent and been offered a wee dram. It seemed quite a normal thing to do as a means of keeping warm. At the end of the events Annette and I presented the prizes to the winners of the Land Rover-sponsored events. Before we left the area we visited Balmoral Castle not far away. Here we met Sandy Masson, the Head Gamekeeper, who I had met a few times before at the Royal Highland Show.

The year 1985 continued to be one in which the company's UK sales were growing and for the month of August, the new registration year and biggest sales month of the year, we achieved record sales of 880 Land Rovers, representing 46.3% of all 4x4 registrations. Range Rover sales of 570 in August were also a record for a single month, but represented just 0.15% of all car registrations. That's how we measured Range Rover performance then.

Land Rover Ltd's growing business was becoming more important to the UK dealers especially as the franchise was becoming more profitable. I continued with John Cunnane's support to manage vehicle orders and vehicle supply so that we could meet customer demand without long waiting times. Keeping supply and demand balanced not only meant dealers did not need to discount the product to make sales, but this approach also meant dealers could enjoy a profitable used-car market for Land Rovers and Range Rovers.

On 15th September 1985 Annette and I were invited by Alan Clark, the Chairman of the Colliers Motor Group, to his company's hospitality suite at the Belfry on the last day of the Ryder Cup. It was a memorable afternoon for me because the Colliers suite directly overlooked the 18th green. Many of the matches were decided before the players reached the 18th but on this afternoon we were fortunate and delighted to witness at close range Sam Torrance hole his 22-foot putt to win his match and so confirm Europe would be victors over the USA team. The final score was Europe sixteen and a half points to the USA's eleven and a half points and meant the USA had lost the Ryder Cup for the first time since 1957.

In the first days of October we launched the latest Range Rover model (code name Aquila) in five regional events to the UK dealers. The invitation said the launch would be held at the Moat House Hotel, Northampton. Dinner that evening would be black tie.

However, we didn't tell the dealers we were going out for dinner. From the front of the coach as it left the hotel I told them, 'Tonight we have something special for you. We are heading to a prestigious location in the country. Our host and hostess tonight are Earl Spencer and his wife Raine, Countess Spencer. So I ask for your best behaviour.'

The venue for our dinner was now known to everyone; it was Althorp House.

I remember standing just inside the main door entrance alongside the welcoming party of Earl Spencer, Countess Spencer, Managing Director Tony Gilroy and John Cunnane, UK Sales Director. Among the first to enter was a dealer unable to shake hands because he had a bottle of whiskey in one hand and a bottle of Coca-Cola under the other arm. This was not what you would call a grand entrance and it took us by surprise.

At these events it was usual for John Cunnane to make an after-dinner speech. John was a brilliant after-dinner entertainer and had a rare talent for a car industry director. His repertoire included amusing stories, anecdotes and jokes and the dealers loved it and were invariably in fits of laughter. On this occasion John stood up and started telling his stories and immediately behind him was a very large vase, Ming, I think. John had a bad back and his body started swaying, and the big vase was in some serious danger. A number of us including me noticed this but John was in full flow and hadn't noticed. Fortunately the vase was still standing at the end of John's speech. On a separate occasion the company had hired a famous comedian to do the after-dinner entertainment. John Cunnane preceded this comedian with his usual brilliant performance and then introduced the comedian, whose first words were 'I can't follow that!' That's how good John was.

I remember the five events at Althorp House because I got to talk to Earl Spencer each evening. I thought he was a kind and very pleasant man. The father of Diana, Princess of Wales, he was now in the spotlight following her marriage to Prince Charles four years earlier. On the evening of 7th October he signed his book *The Spencers on Spas* for Annette, Gillian and me.

November 1985 included a number of Land Rover styling reviews at Drayton Road, which I simply noted in my diary as reviews on 'fascias, stripes and wheels'. None of this was very exciting but every attempt was made to update and refresh the Land Rover products from time to time, even if the basic body shell and styling could not be changed.

The Scottish Motor Show took place in the last part of November at the new Scottish Exhibition Centre in Glasgow and I flew from Birmingham to be there for a couple of days and to attend the Scottish Motor Trade Association's Banquet at the Albany Hotel as a guest of the Taggarts Motor Group. I then flew back to Birmingham for some meetings. However, on 27th November I had to fly back to Glasgow as there was going to be a royal visit to the Land Rover stand. John Cunnane was away, having had surgery on his bad back, and so I would represent the company on our stand that day and greet the royal visitors.

I was told that I should be aware that a Land Rover vehicle being built for the Duke of Edinburgh was not going to be delivered on time and I was warned to be prepared for this to be raised when he visited the stand. When the royal party arrived on the stand I immediately welcomed the Duke of Edinburgh and ushered him towards a Land Rover 110 station wagon he wanted to see. I opened a rear door and he just said, 'I see there is still no b****y room in the back.' There wasn't much I could say in reply other than to agree with him, as we immediately moved on to look at a Range Rover. All too quickly the royal visit was over and thankfully the question of the Duke's delayed vehicle in production was not raised by him. I have a photograph of the Duke of Edinburgh and me alongside the Land Rover as a nice reminder of his visit.

December 1985 was marked by the completion of the UK Sales Operations team's larger refurbished office area closer to the Engineering Block and the Lode Lane main entrance. We all moved in to the new office on the first day back to work on 2nd January 1986.

Franchising activity continued in 1986 with the objective of streamlining and strengthening the UK dealer network. I will use Stratford-upon-Avon as an example. The Heron Motor Group had been the Land Rover dealer there but when Heron sold the property to Sainsburys to be developed as a supermarket we quickly needed to consider our options for a replacement dealer. One day in February 1986 I visited the premises of Yarnolds of Wootton Wawen, a Saab dealer just north of Stratford-upon-Avon, and I asked the owner John Yarnold if he would be interested in taking on the Land Rover franchise. It emerged that he was in the process of building new premises for Saab in Stratford-upon-Avon, and he agreed to represent Land Rover at Wootton Wawen after the relocation of Saab into its new facilities. Although the premises were not ideal this appointment, which took place later in the year, turned out to be the best option available at the time and gave us exclusive dealer representation in an important location.

Communication with our dealers continued to be an important part of our strategy to develop the UK Land Rover dealer network and in April 1986 we held another series of conferences with the five regions. Regional meetings and conferences worked well and as we had a relatively small number of around twenty-five to thirty dealers in each region this enabled us to invite dealer principals and their first-line sales and after-sales management.

On this occasion I drove to Airth Castle near Stirling for the first conference in Scotland. The subjects to be covered at these conferences included an update on the Business Builder programme, which would include an enhanced franchise standards programme, both to be introduced together in January 1987. Also to be discussed were details of the new Range Rover Turbo Diesel (code name Kestrel), service and warranty. Finally, we would hold an election to appoint two dealers to represent the Scottish region on the dealer council.

Following the Scottish meeting I travelled south to the Post House, Wakefield, for the Northern region conference which saw over a hundred dealer staff attend. A similar number were at the Midlands conference at Stoneleigh Abbey at the end of that week. The Southern region was next at the Runnymede Hotel, Egham, and we finished the conference tour at Gloucester for the Western region.

John Sewell was now the Sales and Marketing Director and together with John Cunnane, who had returned early in 1986 from his back surgery, we had regular meetings to discuss production requirements for the UK. It was important that Land Rover senior management recognized the need to take account of what the UK sales team required in the way of production as opposed to my time in BL Cars, when top management decided what the sales team would be required to order to meet fairly fixed vehicle production schedules.

This sales-based management approach enabled us to keep the UK supply and demand of Land Rover vehicles in good balance and so avoid the need for the dealers to resort to discounting and distress marketing to achieve new vehicle sales targets. In addition, this production requirements procedure greatly helped dealers maintain healthy profitability on new vehicle sales and strong margins on their used-vehicle business.

Simon Budd joined the sales department as a second Fleet Sales Manager to support Ron Parr, who for some time had been the sole Fleet Sales Manager. There was one occasion which brought much amusement in our team. At one of my monthly sales meetings Simon was late arriving and so I started without him. When he finally arrived he said he was late because the Land Rover he had driven up the M1 was very slow. It emerged later that he had driven the vehicle up the M1 in the Land Rover's low-range gears, which are usually only used in off-road conditions or where there are very steep gradients. It was no wonder the vehicle didn't go fast, and as a result Simon was given the title 'The Low Ranger'.

August 1986 was a busy month for what I would call above-average events. During that month I met Trevor Finn of Williams Motor Holdings for the first time at his Jaguar dealership, Paramount in Derby. Trevor was responsible for just three or four dealers, including BMW in Hull and Doncaster, and he was interested in taking on the Land Rover franchise in Nottingham. I mention it because some thirty years later he was Chief Executive Officer of Pendragon plc, which he had led to become the largest dealer group in the UK, operating in over 200 locations. However it was not until June 1989 that we were finally able to appoint Trevor's company in Nottingham as a Land Rover dealer alongside its Merlin Porsche dealership on condition that an exclusive Land Rover operation would be established in the city.

Later in August I flew from Birmingham with Tony Gilroy on a day trip to Aberdeen, where he performed the official opening of the Aberdeen Land Rover Centre. I have a photograph of the dealer's directors Bill Broomfield and Donald McHardy, as well as Tony and myself standing outside the new Land Rover showroom. I remember the day well as it was the first time I alone accompanied Tony on a dealer opening.

The following day Annette, Gilly and I set off early for the long drive north to Scotland. We were going to attend the Lonach Highland Games at Strathdon for the second time. Land Rover once again was sponsoring some of the events and we were presenting prizes to the winners. This time we stayed a couple of nights at Kildrummy Castle Hotel, which was built on the estate overlooking the ruins of Kildrummy Castle.

That weekend we had an unusual encounter. While driving on one of the single-track roads high above Balmoral we saw a small convoy of Range Rovers and Land Rovers coming towards us. We instantly recognized the Queen driving the first Range Rover and so quickly pulled over off the road to allow her to continue without stopping. We noticed the Duke of Edinburgh was driving the second vehicle. Other members of the Royal Family including the Queen Mother and Prince Charles were in the two vehicles. I don't know if they were surprised to meet two Range Rovers coming towards them, but it was a surprise for us to see them.

The start of September 1986 marked the arrival of two new university graduates into the UK Sales Operations department for six months until the end of February 1987. I noted in my diary that one was Andy Bruce. Andy was a talented young man who three years later I would promote to be the Regional Sales Manager for the

Scottish region. Much later he became Chief Executive of Lookers plc, one of the largest motor dealer groups in the UK.

At the end of September a showroom promotion was held at John Yarnold's dealership in Wootton Wawen, to mark the appointment of his dealership, now named Arden Land Rover Centre, as the new Land Rover franchise holder for the Stratford-upon-Avon marketing area.

As a brand new franchise it took some time for the business to grow. Several years later, when Land Rover also had the important Discovery in its model range, John Yarnold's Land Rover business had grown to the point where it had overtaken the Saab business. As a result the Land Rover franchise was relocated to the Stratford-upon-Avon premises as an exclusive franchise operation, replacing Saab. This was an excellent example of a family business making a success of the Land Rover franchise opportunity when initially the opportunity was relatively small and success was not certain or guaranteed.

The National Motor Show took place in October and on press day the public launch of the Land Rover Turbo Diesel (code name Falcon) took place. I thought it was interesting that new Land Rover vehicles being developed were often given the code name of a bird of prey, as in the case of the Range Rover Turbo Diesel (code name Kestrel) which was launched to the public at the end of November 1986.

A very busy 1986 was going to stay busy right until the end of the year. In November we held another round of Regional Dealer Conferences to present in detail the Land Rover Business Builder franchise development programme, which included a comprehensive minimum standards programme. Business Builder, which would start in January 1987, was one of the most important operational initiatives introduced by the company and to ensure all dealer principals and their management attended we held two conferences in each region during the period 17th to 21st November.

That month I had a franchise review meeting with Trevor Key and Tony Austin of the Austin-Rover Group at Canley. These would become regular meetings to ensure we were all aware of each company's franchising activities and that as far as possible our individual company's actions didn't have an adverse impact on the other. This was not always possible because Land Rover's overall strategy was to develop an exclusive network where possible and this on occasion did mean the termination of a dealer whose remaining Austin-Rover business alone was less profitable and, in some cases, not viable.

In the 1980s a high level of my communication with Land Rover dealers and customers was in the form of letters, and much of my internal communication to my team and Land Rover management was in the form of memorandums. So each day I was in the office a fair amount of my time was spent dictating responses to letters and internal memos to my secretary, who would then type out what I had said. The main other means of communication were telephone calls or face-to-face meetings.

At that time I often wanted to speak during the day to one of my Regional Managers and having their daily itineraries I would phone a dealer at a time when I expected them to be there. It was amazing how many times they had not yet arrived, or had just left. When I was a sales supervisor in the early 1970s in the Manchester area I thought it was good to be able to visit my dealers and do my business with them without many interruptions from my regional management. The personal mobile phones that we take for granted today had not arrived and would be many years away from being a necessary and unavoidable communication device. I am very glad most of my operational sales career occurred without them.

The days of personal computers for each person to do his or her own electronic mail, most commonly referred to as email, had not yet arrived. I think it was not until around 1995 that I received a big desk computer and learned how to use it. I recall my first email address had CompuServe in it. In 1987 I received a car phone that was a very large piece of kit that was firmly fixed in my Range Rover in the space between the two front seats. It was in practice a telephone similar to a fixed telephone used at home.

In 1987 the Business Builder programme that had taken much time to develop became a major feature in our work with the dealer network. The Regional Managers carried out quarterly audits of each of their dealers and the results were then carefully reviewed to ensure a consistent approach across the UK and that the audits would stand scrutiny if challenged. This was especially important where the dealer groups were concerned because John Cunnane and I realized we had to present the Business Builder audit results to the dealer group heads so that they were fully aware of the financial penalties if franchise standards were not met by their dealers. These dealer group reviews continued to cover sales performance but we now had the extra subjects of franchise standards achievement, potential financial penalties, as well as discussions on an increasing requirement for the dealer groups to invest in the Land Rover franchise in order that we would move to an exclusive franchise network.

An important organizational change occurred during 1987. John Russell arrived from Peugeot to become Land Rover's Marketing Director. John would quickly have a major influence in developing the Land Rover brand, and in introducing a new modern Land Rover corporate identity which included refreshing the Land Rover company logo into the one still in existence today.

Incidentally John added to the already relatively big list of Land Rover's top management with the name John. These included John A. Gilroy, who preferred to be called Tony, John Sewell, the Sales and Marketing Director, John Cunnane, John A. Stephenson, who preferred to be called Alex, John Stephenson in Land Rover Parts, and me. Even more confusing at times was the fact that a few of our initials were the same. So when internal memos were copied to 'JS' or even 'JAS' I often received a memo not meant for me. We were all born in an era when John was one of the most popular boy's names in the UK.

In April 1987 after many weeks of drafting John Cunnane and I produced and circulated a very comprehensive Land Rover UK franchise strategy document to Tony Gilroy and his Directors. It would enable us all to have a clear and consistent understanding and approach to our franchising activity over the coming months and years. We were already communicating our wish to develop a dedicated and exclusive dealer network in the UK. This strategy document explained why we were doing it and the resulting benefits to the company, our dealers and our customers.

In June 1987 we were informed by top management that A. T. Kearney, a firm of Management Consultants, had been appointed to carry out a major review of the whole Land Rover Ltd organization. This review, which was instigated by Tony Gilroy, became a big part of my daily work and that of my colleagues and it continued for several months. A. T. Kearney's particular organization 'zero base' review assumed a starting point of zero. Quite simply I had to justify every person in the UK Sales department team and the associated department budgets.

For example, the role of a Regional Sales Manager was reviewed in detail in terms of cost-benefit. Why did we need Regional Sales Managers, what was their purpose and contribution, and at what cost? I remember many numbered spreadsheets had to be completed within firm deadlines. The end result for my UK Sales Operations department was that we had justified our existence and the organization structure we had in place. I can't recall any major organization changes resulting from the A. T. Kearney review elsewhere in Land Rover but I remember the detailed work my team and I had to do to justify our individual and collective existence.

During the Royal Show at Stoneleigh in July we had our now usual VIP visits to our stand and this year we lined up to welcome and meet Princess Anne. During the summer I played an occasional round of golf with John Cunnane at his golf club in Kenilworth and I was taking an increasing interest in this sport even though I had no official handicap and did not belong to any golf club.

One event affected me directly on the night of 15th October 1987. I had travelled to the Colchester area to meet the directors of the Lancaster Motor Group for dinner at the Dedham Vale Hotel. The plan was for us to meet again the following morning at their Head Office in Colchester to discuss ways in which the Lancaster Motor Group could take on the Land Rover franchise. Later that night there was a big storm which woke me up as I could hear the sound of tree branches breaking and crashing outside. Early in the morning after some disturbed sleep I woke up to find everything in darkness and a power cut. The sight outside was hard to describe. Large trees had come down during the night and there were tree branches everywhere. It was a site of devastation I had never seen before.

I had parked my Range Rover directly underneath one of the few large trees that didn't come down in the hotel's grounds and as it was a Range Rover I did manage to drive over tree branches and eventually leave the hotel. In an ordinary car that would have been impossible. On arriving at Lancaster's Head Office I was told the directors could not get to meet me. So my business meeting did not take place. On my journey back home it was clear this 'Great Storm of 1987' had affected large parts of eastern England and the East Midlands as no petrol stations were open owing to local power cuts. Fortunately I had enough fuel to get home.

A little over a month later on the afternoon of 24th November 1987 Land Rover held one of its more dramatic and unusual conferences for the UK dealer network at the Sheraton Skyline Hotel, near London Heathrow Airport. The conference was called 'Land Rover Live' and was recorded by a film camera team. Andrew Davies of Cricket Communications who organized the conference proposed a format where each of Land Rover's top managers would be interviewed in turn by Richard Kershaw, a well-known television reporter and interviewer who had worked on the BBC's *Panorama* television programme. A setting was created whereby the audience, as in a live TV programme, would comprise the several hundred members of the dealer network.

Rehearsals were held the day before to discuss the form of a business interview, with questions from Richard and answers from me. Richard told me he would reserve the right to invite members of the audience to challenge what I had said. My colleagues Chris Langton and Allen Murdoch would speak on behalf of service and parts functions respectively.

There were two important parts or messages in my 'interview' with Richard Kershaw. First I had to explain the results of the first Business Builder audits during 1987 and to tell the dealers what we were proposing to do with the moneys that we had withheld from dealers that had failed to comply with the franchise standards. The policy to withhold 3% of the dealers' Land Rover margin and 3.5% of the Range Rover margin was not welcomed. Richard was a formidable interviewer and true to his word asked the audience, 'Do you agree with what he said?' There were some mumblings from some of the dealers but fortunately I received no strong heckling or verbal abuse.

Second, I had to tell the dealers about the next stage of our franchising strategy. I explained that in two years' time we would be introducing an important third new model which would sit between the existing Land Rover and Range Rover models. This new model was already the subject of much dealer and media speculation as Land Rover needed a vehicle to compete with Japanese models in the growing 4x4 sector.

Our franchise strategy and objective would be to have an exclusive dealer, wherever possible, in place by November 1989 to exploit the potential of this additional model. This vehicle while under development in Land Rover was called 'Project Jay' and when launched would be called the Land Rover Discovery. However, this strategy would mean a reduction in the dealer numbers and I recall saying that if you look at the dealer either side of you, one of you won't be here in two years' time. In practice this meant we were aiming to have a network of some 127 dealers in time for the new Discovery launch.

It was a good experience to be interviewed rather than do the conventional presentation. In addition, as the conference was filmed a VHS recording was given to some of us. I still have my VHS tape recording of the event. The 'Land Rover Live' conference ended with a speech from Tony Gilroy, the one person among the Land Rover 'presenters' to avoid an interview with Richard Kershaw.

That evening our guest after-dinner speaker was Captain Brian Walpole, who led the Concorde fleet for British Airways from the beginning of the supersonic era and was the mastermind of its commercial success. That night at the end of what I considered to be a successful conference I retired to my room. A few hours later I was woken up by the room phone ringing. I was shocked to be told one of the dealers was floating in the hotel's swimming pool on the ground floor. I got dressed and went down to the swimming pool and fortunately Noel Calvert the dealer concerned was alive and well. He had however been seen just in his underpants floating rather than swimming. I didn't see it but if you knew Noel what the hotel staff saw in the swimming pool couldn't have been a pretty sight.

That wasn't the end of the conference story. In the morning I asked for my bill at the reception desk to check out and I was horrified to see a bar bill of many hundreds of pounds. The previous day the dealers had been told that drinks would be paid for by Land Rover until 11 p.m., but after that the dealers would have to pay for more drinks. However, the dealers had decided Land Rover should continue to pay and having obtained my room number decided to put the cost of more drinks on it. I was not best pleased but I had no option but to accept the additional cost and put it on the main conference account.

The year 1987 turned out to be a very successful one for new Land Rover sales in the UK. In January 1988 I wrote a letter to the dealers with the sales performance for the full year 1987 stating that Land Rover 90 and 110 registrations were 6,243 against an objective of 6,000, while Range Rover registrations were 5,027 against an objective of 4,200. It was the best ever year's sales of Range Rover in the UK.

At that time August was the biggest month for vehicle registrations as it was the month when the vehicle's registration year changed and a new registration number appeared. So in August 1987 the 'D' prefix changed to 'E'. The Range Rover objective for August 1987 was 800 registrations and having added up the Regional Sales Managers forecasts I was confident the objective would be achieved. However, the final figure was 799 registrations, and so while it was in fact a record month's sales for Range Rover I had failed to achieve the objective by one vehicle.

I am using polite language here instead of Tony Gilroy's actual non-polite words. He gave me a very strong reprimand and deservedly so. He told me I had the ability to register a number of company vehicles to be sure of achieving the Range Rover sales target but I didn't do so. I had to accept that the publicity benefit of saying we achieved a record 800 Range Rover registrations would have been much better than the 799 we actually did.

The year 1988 started with an invitation from Joe Nimmo, Managing Director of the Taggarts Motor Group, to attend the 'Burns supper' on 25th January at Ibrox Stadium, home of the Glasgow Rangers Football Club. I remember we were sat on a table with John Greig, the Rangers captain, and the dinner was preceded by the traditional 'address to the haggis'. This was in fact a poem written by Robert Burns to celebrate his appreciation of the haggis.

As a result Burns and haggis have been forever linked. This particular poem is always the first item on the programme of Burns suppers. The haggis is generally carried in on a silver salver at the start of the proceedings. As it is brought to the table a piper plays a suitable, rousing accompaniment. An invited artist then recites the poem before the theatrical cutting of the haggis with the ceremonial knife.

From 14th to 19th February 1988 along with other senior Land Rover managers I started a management diploma course at the

Cranfield School of Management, near Bedford. During the first part of the course we were involved in a wide range of 'business games'. The second part took place in the last week of April and included a management task outside. The task required us in two teams to cross an imaginary river about ten yards in width, without touching the imaginary water. To achieve this we were given some material which included some scaffolding, wooden planks and ropes. Ken James climbed up to the top of some scaffolding we had erected but got stuck for a while, being unable to move forward or back. Unfortunately his predicament seemed rather comical and produced some laughter from us which didn't amuse him.

At the end of the second week each of us had to make a presentation to Tony Gilroy, Alan Curtis, the HR Director, and Alan Simpson, Managing Director of Land Rover Parts. This presentation had to include recommendations on how to improve Land Rover's business, and we were informed that we should refer to some theory and practice that we had learned at Cranfield. Most of us were concerned that some of our possible suggestions would appear to be too critical of the top management and so some of the content in our presentations was watered down to avoid being too contentious.

On 24th April 1988, the Littlewoods League Cup Final took place at Wembley. I mention it because Luton Town, the football team I had supported from an early age, beat Arsenal 3–2. Luton stunned the defending champions and the competition favourites Arsenal. This was the first major trophy that Luton Town had won in its history.

In June I travelled once again to the Royal Highland Show in Scotland. In July I attended the Royal Welsh Show at Builth Wells. At the latter we stayed at the Kilvert Hotel in Hay-on-Wye, where in a similar tradition to the Scottish Show we hosted a dinner for the Welsh Land Rover Dealers. A few of us also found time to play a round of golf with Mike Like, owner of the Land Rover dealer J. V. Like & Sons of Hay-on-Wye. This round took place at the Cradoc Golf Club, near Brecon. I remember it well because one of the holes was played up one of the steepest slopes I have ever climbed. I think it was known locally as Heart-Attack Hill.

In August the UK Sales Finance department under Barrie Wood was added to my existing UK Sales Operations team responsibilities. Barrie's department was renamed UK Dealer Business Operations and its aim was to strengthen support to the dealer network. Barrie's broader responsibilities now covered dealer agreements, dealer stock management and financing, dealer viability appraisal and monitoring as well as sales programme administration, including Business Builder. I was already very involved in sales finance matters, working closely with Barrie and his team, especially on day-to-day operational issues related to the cost of funding vehicle stocks in the dealer network.

During that month I attended some new product launch meetings, both with names of birds of prey. One was Osprey (the new Range Rover Vogue SE), the other was Project Jay. Jay was a very important project for Land Rover as it was to create a model range that sat between the Range Rover and traditional Land Rover, to fight the growing competition from Japanese models. Meetings on Project Jay had now become a regular feature in my diary, having commenced a year earlier. August 1988 also represented the 40th anniversary of the introduction of the Land Rover and I attended a number of dealer showroom promotions to mark the occasion.

Throughout 1988 I was increasingly involved in reviewing proposals and receiving presentations from many dealers relating to their planned investments in exclusive Land Rover premises to meet the company's specific franchise requirements. Many dealers, including those within national dealer groups, had been given a two-year period to make proposals and if approved implement their plans by the proposed launch of Jay in November 1989. By September 1988 time was running out for some dealer plans to be approved and implemented.

In October 1988 I attended another Land Rover closed auction, at the British Car Auction premises at Measham. The auctions were 'closed' in that only official Land Rover dealers were invited to attend. We wanted to make sure only official dealers were able to source and purchase Land Rover Ltd's company-used vehicles and to take advantage of the profit opportunity these provided. However, occasionally non-franchised traders tried to gain entrance and so a few Regional Managers were in attendance to spot any intruders.

These vehicle auctions were generally held when we had around eighty to a hundred vehicles to sell. These vehicles often included the Range Rover I had recently replaced with a new one and there was usually some muttering from the dealers when the auctioneer announced the next Range Rover to be auctioned was a carefully driven senior manager's vehicle. Mine!

George Hassall who was responsible for Land Rover's company vehicles would stand alongside the auctioneer in order to 'communicate' whether we would sell a vehicle that had not reached its reserve price. This 'communication' invariably took the form of a slight tug from George's hand on the auctioneer's jacket, so that the assembled dealers were not aware of a reserve price not being achieved but where we were happy for the vehicle to be sold. The last thing we wanted to do was to take unsold vehicles back to the factory.

The UK dealers had achieved record sales from relatively low vehicle stock availability and at the end of October 1988 there were just 567 Land Rovers and 441 Range Rovers physically held in stock in the UK dealer network. This relatively low vehicle availability coupled with growing customer demand enabled dealers to achieve a very good level of profitability from new sales, while profitability from used Land Rovers and Range Rovers was also very strong.

Towards the end of the year there was a somewhat unexpected change at the top of the Land Rover organization. On 7th December I was on a two-day business visit to Ireland, and while there I was surprised to learn that Tony Gilroy, Land Rover's Managing Director, was leaving the company.

Chapter 10

# Land Rover Ltd 1989

**ROVER GROUP'S CHAIRMAN, SIR GRAHAM DAY,** appointed George Simpson in January 1989 as the new Rover Group Managing Director. Graham arrived in 1987 when Rover Group was privatized in a sale to British Aerospace (BAe). George had previously been Managing Director of Freight Rover, the maker of light commercial vans. On becoming Managing Director George almost immediately reorganized the company, replacing the three boards of Austin-Rover, Land Rover and the Rover Group with one single board. Until then the management in Land Rover assumed that the plan was for Land Rover to be privatized as a separate independent company. That was not going to happen now.

Tony Gilroy's leaving event took place at the Land Rover Social Club in Solihull on 31st January 1989. Tony had a strong positive impact on me at Land Rover, and I know through my dealings with him that he made me a better manager. He would often phone me for some information. I quickly learned that he wanted facts not opinions and that if I didn't know I told him I would call him back as soon as I had the facts or the answers he wanted. Guessing or waffling was not the way to respond to Tony's questions as some found out to their cost.

It was obvious to all who worked in Land Rover Ltd, even if they did not have direct contact or dealings with him, that Tony had a strong competitive personality. That was evident at his leaving 'do'. Tony and I had a game of snooker and even though he needed a number of improbable snookers to beat me he refused to give in and accept defeat until it was clearly impossible for him to win.

The 1989 Business Builder programme included a strong message from Tony on its first page. He stated that: 'The ability to fully satisfy the customers' needs and expectations must be constantly addressed. Range Rover has firmly established itself as one of the most desirable luxury vehicles in the UK and the discerning customer rightly expects the highest level of care and attention. We must strive to ensure that all aspects of the Land Rover franchise are as good as, if not better than our competitors. Therefore, the 1989 Business Builder programme incorporates an important new development which will allow, for the first time, your individual performance as a franchised dealer to be assessed by your most important assets – your customers.'

In the 'Introduction' section it was explained that 1989 Business Builder Standards comprised two elements. The first were 'Franchise Standards', which as minimum standards had to be met in full to gain entry into Business Builder in addition to being a condition of holding the franchise. The second were 'Operating Standards', designed to improve the quality and professionalism of the dealer network. Both sets of standards were grouped into Sales, Service and Parts and were subject to quarterly assessment.

The programme generated its funding by retaining a percentage of the dealer discount, 3% from most Land Rover vehicles and 3.5% from Range Rovers. This meant net dealer discount was 13% on Land Rovers and 15% on Range Rovers and Ninety station wagons. One hundred points were available if a dealer achieved 100% of Operating Standards. Rebates were paid at the end of each three-month period to reflect the number of points awarded following the dealer's quarterly self-audit. We had introduced the process of a dealer self-audit for two main reasons. The first was because we wanted the dealer management to take a greater ownership of the programme and to measure themselves on all the key parts of their business. The second reason was because it would reduce the amount of time the Regional Sales,

Service and Parts teams spent on auditing their dealers and the associated administration.

From July 1989, however, 25% of the points would be subject to the new Customer Satisfaction Rating Index (CSRI) which would result from independent surveys conducted among the dealers' new vehicle customers. The introduction of customer satisfaction measurement as a major factor in awarding dealer rebates was controversial with many dealers at the time as they felt this added too much of a subjective customer factor into the process.

To achieve 100 points on the 1989 Business Builder programme was difficult and challenging for many dealers, especially those whose management were not totally committed to the Land Rover franchise. To ensure there was a strong franchise network discipline in operating the programme we also introduced penalty points for major breaches, such as fraudulent claims and no claim submission. However, when assessing the dealers' self-audits I can remember briefing my Regional Sales Managers to give the dealer the benefit of the doubt if there was genuinely a 'grey area' in its claim.

In February 1989 we held a further series of regional dealer conferences. John Cunnane and I had a number of meetings with Andrew Davies of Cricket Communications to discuss and rehearse for these conferences which were held in Livingston, Scotland; Leeds; Egham, Surrey, Cwmbran, Wales; and Melton Mowbray. Andrew would now work with Land Rover on conference management and delivery and he brought a refreshing, positive and innovative approach to our important communications with the dealer network. He was very influential in deciding the way conferences would be held, in conference script content, and he also made sure we had proper rehearsals for all the company speakers.

During the same month Chris Woodwark took over from Tony Gilroy and a new Land Rover era was underway. In March 1989 some evidence of the new integrated Rover Group structure emerged when I was invited along with other senior managers to a Rover Group Executives' meeting at Canley, Coventry, where we were briefed by Sir Graham Day and George Simpson on the company's plans.

What I remember thinking at the time was this. When I joined the Austin-Morris Division of BL in 1969, it was run separately from Rover-Triumph and Jaguar Daimler. In 1975 the Ryder Plan integrated all the car brands into one Leyland Cars structure. In early 1979 for a short while Sir Michael Edwardes separated the commercial activity of Austin-Morris from JRT, but towards the end of the year he changed his mind and they were reintegrated. In 1984 Jaguar Cars under its Managing Director John Egan was privatized and Land Rover Ltd was running fairly independently in the 1980s and had its own board. Now in 1989 Land Rover was being integrated once again at board level with the Austin-Rover car brands. Our leaders were continually putting all the brands together, taking them all apart, putting them all together etc., etc. Consistency of strategy and direction was probably a much better way to run an international automotive company.

Fortunately for Land Rover in the UK we were a long way down the road of implementing a franchise policy that would create an almost totally exclusive and dedicated dealer network – a policy that would be very hard to reverse. In reality it was not a policy we had to change but the new structure did bring extra pressures on us from top Austin-Rover management to keep Land Rover alongside its dealers, when we were requiring the dealers to make investment in exclusive Land Rover premises.

However, I and other senior managers now became very involved in implementing the Rover Group's Total Quality Improvement programme (TQI). This meant I now spent a much larger amount of my time looking inwards into our own company's activities and processes rather than outwards at those external activities with the dealer network that would grow sales and improve the customer experience of the Land

Rover brand. A few months later in May we had to make presentations of our 'personal action plans' to George Simpson.

At the same time in April the Land Rover Commercial team, led by Chris Woodwark, started a process to review and produce a new mission statement for the business. It culminated in June 1989 in a team meeting at Moor Hall in Sutton Coldfield, where twenty-five senior managers formulated the following mission statement: 'We will provide Land Rover customers with outstanding products and an excellent ownership experience through the best distribution service in the world.' We all agreed it was a challenge but possible.

Following this senior management work, I was asked to put together a presentation on the Land Rover Commercial mission, so it could be communicated to everyone in the Commercial team. This I did and the mission was presented by function management as planned.

One memorable Land Rover dealer showroom opening I attended took place in April 1989 at Westover Motors in Poole, Dorset, where the Managing Director was Peter Wood. I remember Nigel Mansell's red Ferrari F40 was on display and there was an auction at the end of the evening. The star item being auctioned was the driver's helmet that Nigel Mansell had worn in winning one of his Formula One British Grand Prix. There were two gentlemen who were very keen to acquire Nigel's helmet and each time one thought he had made the final bid, the other increased the bid. This outbidding process between the two men went on like this for some time until eventually it was auctioned for a very considerable sum. When I asked the winning bidder about the size of his bid he just said, 'When Nigel Mansell dies it will be worth a lot more.' However, as I write this Nigel is still very much alive.

In May 1989 we travelled for our family holiday to Acapulco, on the west coast of Mexico, where we stayed for two weeks at the Fiesta Americana Condesa Hotel. Gilly had taken a cassette of Bros, a popular British singing brothers' duo at the time, so before long we were all listening to Bros songs by the hotel pool. Towards the end of the holiday, we took a taxi one evening to the western side of Acapulco Bay to see the cliff divers at Mirador La Quebrada, who perform daily shows for the public, diving 35 metres from the cliffs of La Quebrada into the sea below. However, away from the five-star upmarket hotels along Acapulco Bay, 'downtown' Acapulco looked very scruffy and not the place for tourists to go.

A week later I made the customary journey at this time of year to Scotland to attend the Royal Highland Show. On this occasion we also arranged a 'Social Day' for the Scottish region dealers at the Gleneagles Hotel in Auchterarder. The dealers and their partners were offered a choice of sporting activities that included shooting at the Jackie Stewart shooting school, indoor leisure pursuits in the hotel and golf at the famous King's Course.

Golf was now becoming a more frequent and enjoyable activity for me and soon after I was invited to play in Rex Holton's Hacfield Golf Day at the Woburn Golf Club. The Duke's course that we played at Woburn was and still is a championship course and so it presented a real challenge to me. In my diary I noted that I went round in 99 shots and scored 27 points. I was pleased with that as it showed I would probably not disgrace myself on a similarly difficult golf course.

The Woburn course was difficult but I would come back to play there in the Dunhill British Masters Pro-Am in 1990 as well as three Weetabix-sponsored Women's British Open Pro-Am tournaments in 1992, 1993 and 1996.

In the same week I attended an executive's meeting at Canley, where it was announced that the Honda Motor Company had taken a 20% share of the Rover Group. This would have a significant bearing on not only the continuity of the Rover Group but also in the product development of Rover models in the coming years.

In July Annette flew from Heathrow to Warsaw to visit her Polish relatives. That month Land Rover management gathered in the Land Rover Social Club to wish Reg England well in his retirement. During the period 1983 to 1989 while I was UK Sales Operations Manager, Reg was the UK Service Manager, and so we worked closely on franchising, franchise standards, dealer development and the various operational matters affecting sales and service.

I played an occasional round of golf with John Cunnane at his Kenilworth Golf Club. In August 1989 I came very close to getting a hole-in-one, something I have still not achieved. On the par-three 196-yard 10th hole I hit a good tee shot straight towards the green, in front of the clubhouse, but it was not until I reached the green that I saw how close it was to the hole. One of the club's ground staff who was standing behind the green told me the ball rolled towards the hole and said he was sure it was going in but as the ball almost stopped rolling it touched the side of the cup and finished three inches behind the hole.

# Chapter 11
# Land Rover Discovery

**I HAVE ALREADY REFERRED TO 'PROJECT JAY',** which was to develop and introduce a third model into the Land Rover product range. This vehicle was to become the Land Rover Discovery. It was the most important new vehicle I was involved in and so I am going to record my part and experience in the project and in the launch of the vehicle.

The decision to introduce a third model was made not only to bring a new generation of customers into the Land Rover company, but also to address, enter and exploit the growing 'personal transport 4x4 sector' of the vehicle market dominated by Japanese competitor vehicles such as the Isuzu Trooper, Mitsubishi Shogun (Pajero) and Nissan Patrol.

Work on 'Jay' started in around 1986. The broad product plan was to base the new model on the running gear of the Range Rover, but with a simplified 'lifestyle' specification, and a more leisure-based styling to meet the aspirations and expectations of potential customers in this mid-range 4x4 market sector.

I first became involved in the project around April 1987 when it was decided the UK market strategic objective was to have, as far as possible, an exclusive dealer network in place to launch the new model towards the end of 1989 and exploit the new model's potential. At the same time we wanted all the dealers in the network to have the capability to grow the sales of Range Rover and the Land Rover 90 and 110.

During the few years leading up to its launch I was fortunate to see the clay and preproduction Discovery models at their various stages of evolution and I was able to give my views and input from a sales viewpoint. In particular I remember the product 'clinic' where potential customers were shown a preproduction Discovery model stripped and devoid of any reference to Land Rover. Among many questions they were asked to say what they thought the retail price of the vehicle would be if it was a Ford, a Vauxhall, a French brand, a Toyota, a Land Rover and a number of other brands. What emerged from this exercise was a suggested price for the Discovery that had a premium of at least £1,000 over every brand and more in the case of those brands not known to have 4x4 vehicle credentials.

Much of my input also related to the vehicle specification, pricing and initial launch production. The main specification was as follows. Much of the running gear was from the Range Rover. It used the Range Rover's V8 engine and the same manual gearbox, but in order to distance itself from the Range Rover the V8 in the Discovery had twin SU carburettors, whereas the 2.5-litre diesel engine was new to Land Rover.

Project Jay would use a great deal of hardware and components from the Range Rover. This not only had a positive impact on keeping down costs, but it also sped up the vehicle development process so that the Autumn 1989 date for introduction of the model could be achieved. Many of the carry-over parts were highly visible, such as the Range Rover body panels, doors, the windscreen and the Morris Marina door handles. I think the Discovery also had Sherpa van headlamps, Maestro van tail lights with the Austin 'chevron' on their lenses, and the vehicle's ashtray had a BL logo stamped into it. Despite all of this it was still an ambitious vehicle development programme.

It was decided the Discovery would be produced initially only as a three-door vehicle, mainly to protect Range Rover sales and to minimize the possibility of Range Rover customers deciding to trade down to a Discovery. This three-door-only launch would give me and my sales team some operational difficulty because it was clearly more logical to offer potential customers a five-door version of such a vehicle. This was an issue that emerged as expected when only a three-door model was launched. However, the Discovery was still a relatively stylish vehicle for the time with strong visual side graphics and Alpine windows.

The main innovation was the interior. The Conran Design Group gave the Discovery a bright interior with light colours, soft-feel plastics and many customer-friendly features. The majority of the interior was constructed from 'Sonar Blue' plastic, with blue cloth trim, with apertures above the windscreen, handholds for rear passengers incorporated into the head restraints of the front seats, remote radio controls on the instrument cluster, twin removable sunroof panels and a special zip-up storage bag behind the rear seats. There was also a branded cloth fabric holdall in the front centre console, for storing small personal items, and which could be removed. This could be worn as a 'handbag' using a supplied shoulder strap. The interior design was unveiled to critical acclaim, and won a British Design Award in 1989.

Land Rover Ltd's vehicle naming policy also changed around this time. For many years there had been a double use of the words Land Rover. Land Rover was the company name and also the vehicle name. Following the lead of John Russell, on the need to clarify the Land Rover brand, Land Rover would now be the brand name, with Discovery being the model name. The original Land Rover 90 and 110 would now be known as the Defender 90 and 110.

John Russell was instrumental in leading the senior management team to produce Land Rover Marque or Brand Values. These would be six words that would be at the heart of everything Land Rover did. The words would represent the values of Land Rover people and the distinctive values of the Land Rover vehicles and be reflected in all of the company's marketing. After much discussion and debate in the senior management team the final six words selected were:

1. Individualism – Knowing your own mind; independence.
2. Authenticity – Fit for purpose; the original.
3. Freedom – Go where you want to go; choice.

4. Adventure – Exploring the unknown; with care for the environment.
5. Guts – Giving everything you've got; endurance.
6. Supremacy – Superior to all competitors; leadership.

To ensure these Land Rover Marque Values were learned and retained in the staff's mind a handy plastic card of credit-card size was produced describing the six words. I still have my plastic card with the original Land Rover Marque Values. Incidentally I can remember some concern being raised by the Land Rover management in Spain, as the word 'Guts' when translated into Spanish just had the local meaning of 'entrails'.

Following the Range Rover presentations at Althorp earlier in September 1989 I and my senior management colleagues became very involved with Andrew Davies of Cricket Communications and his team in developing presentation scripts for the Discovery launch to the UK dealer network. This launch would take place in two events in Plymouth during the period 13th to 16th October. At the same time I started to prepare my operational plans for the dealer launch of Discovery to their customers. It was my once-in-a-lifetime chance to get the new Discovery launched in style, uniquely, professionally and with a strong, consistent message to the public.

A few months earlier we had to make decisions on the initial Discovery launch stock that the UK dealers would have to demonstrate and sell. This meant determining the mix of petrol and diesel models as the vehicle would be offered with both V8 petrol and 2.5 diesel engines. The three-door Discovery would be offered initially at launch with just three option packs in addition to a base model. The three option packs comprised of an Interior pack, an Exterior pack, which included alloy wheels, and an Electrical pack. However, with my input and that of my like-minded management colleagues we priced the three option packs so that it was better value for a customer to buy two packs rather than one, and three packs gave even better value than buying two. The base Discovery model would have a retail price of £15,750.

I recall having a strong input on this as I felt we needed an optically attractive starting price. I knew you could always put a price up but rarely put one down. However, we knew customer demand would outstrip supply for many months after launch and so while it was on the price list we decided there would be no base models in the dealers' initial launch stock.

Information and details of the Discovery, including photographs of the vehicle, had leaked into the press and dealers were reported to have taken many orders in advance of the launch, without the customers having seen one in the flesh. We decided on the basis of the pent-up demand for this vehicle that the dealer launch stock would have a rich mix of two and three packs. We made it a firm requirement for all UK dealers that one of the first vehicles dispatched to them would have to be notified to us as a demonstrator, then licensed and run as a demonstrator, even though we knew the total of 127 vehicles could all be sold immediately. We felt it was important and essential in order to achieve conquest sales that potential customers driving vehicles of other brands had the opportunity to experience the new Discovery.

I have kept some of the launch documentation that I produced at the time and so I am able to record here accurately how the launch of the Land Rover Discovery evolved in the UK, as well as my report to the company's directors of the launch. While I have some of the paperwork I am now aware the communication exercise at the time in the form of letters to the dealers and internal memos used up a great deal of paper. The electronic communication methods of today would have made what we did much easier, quicker and cheaper.

On 11th September 1989 I sent my first letter on the Land Rover Discovery to the UK main dealers. The first paragraph read as follows.

> *As part of the Discovery launch programme we shall expect each main dealer to hold a promotional event on the evening of 15th*

*November, to which selected potential customers will be invited to view our exciting new vehicle. It is intended this promotion should include brief introductory remarks on Discovery by both the Dealer Principal and a representative from Land Rover.*

The dealers were asked to complete and return a form to me confirming details of their proposed launch event. The following day Land Rover unveiled the Discovery at the Frankfurt Motor Show, at which point the new model was firmly in the public domain.

A short while earlier I had asked Land Rover's Directors to provide me with a list of persons they were willing to nominate from their area to represent Land Rover at a dealer launch event and make a speech on behalf of the company. I received great support on this from Land Rover Directors Terry Morgan, Brian Purves, John Russell, Stephen Schlemmer, Alan Simpson, Chris Langton and Mike Donovan. As a result on 5th October I confirmed to them and Chris Woodwark that they had provided me with over 150 nominations to support the 127 dealers in launching the Discovery. I confirmed to the additional number of extra nominations that while they would not be required to make a formal presentation they would be 'deployed' to support the more significant dealer promotional events.

The launch presentations of the Discovery to the UK dealer principals would take place between 13th and 16th October in Plymouth, a city on the south coast of Devon. Separate presentations to fleet users and dealer staff were scheduled to follow and take place in Plymouth between 4th and 8th November.

Plymouth was chosen for the Discovery launch for a number of reasons. It became famous for the achievements of sea captains based there, such as Francis Drake, who began his sea battle in 1588 off Plymouth which led to the defeat of the Spanish Armada. Many voyages of discovery started in Plymouth, notably in 1620 when the Pilgrim Fathers set sail to the New World of America, where they established

Plymouth Colony. Plymouth was known locally as the city of discovery so it was an appropriate venue to launch our Discovery. The Dartmoor countryside in the area would also be perfect for the ride-and-drive, when the dealers would drive and experience the vehicle for the first time. Land Rover marketing created a conference facility by refurbishing a warehouse on the dockside. From the outside no one would know this was where the Discovery would be presented and launched. The UK location would also be logistically good and relatively cost-effective for a new vehicle launch.

Eighty-six Discoverys were used for the ride-and-drive activity. They all had a 'G' registration plate with consecutive registration numbers and with the last three letters WAC. To fans of the Discovery these vehicles would soon became known as 'G-WACS'. I don't remember the vehicles being called G-WACs but coincidentally I had just taken delivery of a new Range Rover with the registration number G152WAC.

According to local press reports from the time, the ride was soft, and roll angles limited enthusiastic cornering, but off-road ability, tested on Dartmoor, lived up to established Land Rover legend. One of the launch highlights was a trip to the South Devon Railway at Buckfastleigh, on the edge of Dartmoor, where a suitably modified Discovery pulled several railway coaches along the railway track.

There were two dealer launches, the first for the Scottish, North and Midlands regions, and the second for the South and Western regions. All aspects of the launches went well. At the end of the dinners Sir Graham Day, the Rover Group Chairman, made a speech and following this the dealer principals and their partners left the dinner tables for some after-dinner drinks.

On 24th October I sent a memo to all the persons who would be representing Land Rover at a dealership Discovery launch on 15th November. In this memo I confirmed details of the dealer each person would be supporting and that on 6th November Andrew Davies and his team from Cricket Communications would be holding a briefing session to guide everyone on the presentation format. I concluded the memo by saying, 'On the night of 15th November we have a unique opportunity to get across a powerful message on Discovery to some 20,000 potential customers. We cannot miss the opportunity to obtain immediate customer reaction to Discovery and therefore I am asking each person to complete a response form and return to me. This will quickly give us a strong national picture following launch night.'

I wanted to know how many people attended the launch, customer reaction to the Discovery, including their likes and dislikes, and any other relevant comments. On the same day I wrote a letter to all Land Rover main dealers confirming the name and job title of the person who would be representing Land Rover at their launch event.

While this was going on, Andrew Davies of Cricket Communications had produced presentation scripts for the dealer principal and the Land Rover representative to make so that the Discovery launch communication to potential customers would be clear and consistent. In addition, and to leave nothing to chance, combined presentation training and launch briefings sessions were arranged for dealer principals in the week commencing 30th October. On 6th November a similar briefing session was held for the Land Rover representatives taking part in the dealer launches. Our objective was to do everything we could to ensure professional presentations in launching the Discovery were going to be made throughout the dealer network on 15th November.

On 4th November I returned to Plymouth for a presentation of the Discovery to selected fleet users to be followed by presentations to three groups of UK dealer staff. For these groups I made a sales presentation which I remember mainly because of some unexpected trouble I had with the autocue system on one of the days.

Normally autocue is a straightforward and very helpful device for presenters and the system is still used today, for example, by television newsreaders. However, as soon as I started my presentation I saw that a fly had got into the machinery and was trapped there with the result that while the words were scrolling forward the fly was walking around the words I was meant to be reading. It was very distracting to say the least. However, I immediately knew there was no point in me referring to this 'fly in the autocue' distraction because it would not make much sense to most in the audience and I would probably lose my focus on my speech.

This was an exceptionally busy period for me and for my Land Rover colleagues, and the main Discovery launch event by the Land Rover dealers to their customers and potential customers was still to come. By week commencing 13th November almost all the dealers had submitted their Discovery launch plans for the evening of 15th November, because a few had decided to do it on 16th November, the public launch date. Then I received a couple of surprises.

The first required a late change of plan for me regarding the dealer I was going to support on the 15th November Discovery launch. I received a phone call from the Rover Group Chairman's PA telling me that Sir Graham Day wanted me to accompany him on his visit to Appleyard of Leeds, where in addition to launching the Discovery, he would be opening Appleyard's new exclusive Land Rover facilities in the city.

The second surprise came after I received a phone call on 13th November from John Russell, Land Rover's Sales and Marketing Director, saying he wanted to see me in his office the following morning. When I got there he explained that John Cunnane would be leaving the company and that I would be offered the position of Director, UK Sales in Land Rover. I was genuinely surprised at the timing of my promotion even if one day in the future I hoped to take over from John. I was naturally happy and excited about the challenge shortly to face me as a Land Rover Director but I was told I had to keep this news to myself for a while until official announcements were made by the company.

On 15th November I drove north to the Wood Hall Hotel near Wetherby in Yorkshire where I met Sir Graham Day and his wife Anne. We discussed the Discovery launch, the speech he was going to make, and I briefed him on Land Rover issues that I thought may arise in the interview which he was having with the local press after the launch. He was keen to learn from me what questions I thought he might be asked and in some cases how he should respond. I found him very easy to talk to and during our conversation he even had time to tell me about his early career as a shoe salesman in Canada.

That evening at 7.30 p.m. on the premises of Appleyard of Leeds the Discovery was duly launched and the new Land Rover facilities officially opened. Appleyard Directors Mike Williamson, Chris Welch and Robert Hirst represented the Appleyard team. Sir Graham Day and I represented Land Rover. It emerged I was the only Land Rover executive scheduled to speak and support the Discovery launch at a Land Rover dealer on 15th November who did not speak that evening, but I was delighted to support Sir Graham Day.

All UK dealerships launching the Discovery were supported by Land Rover management, demonstrating Land Rover's commitment to the Discovery and to our dealers and customers.

I still have the original presentation scripts and so before they disappear I am reproducing the dealer principal's speech script followed by that of the Land Rover representative. Almost word for word, this is what they had to say.

> *Ladies and gentlemen, good evening and a very warm welcome to you all. I'm delighted you have been able to join us at (dealer name) for the launch night celebration of the new Discovery from Land Rover. Welcome to our existing and valued customers. Welcome also to a great many new faces because more than half of you I know may not have been Land Rover customers before. [Optional section if opening new facilities.] Tomorrow marks the official public launch of Discovery – the first all-new vehicle from Land Rover for nearly twenty years. We're very excited about it and so we wanted to take this opportunity to hold a privileged preview before an invited audience. Discovery has been comprehensively designed and*

*developed to fit into a sector of the UK car market – one in which we haven't previously competed. Discovery offers customers like yourselves a multipurpose flexible vehicle for both business and leisure. First and foremost it's an on-road estate car, but it offers the go-anywhere flexibility of a four-wheel-drive vehicle with a sophisticated car-like exterior design; superbly integrated interior styling; best-in-class performance; and exceptional value for money.*

*On the exterior it's a vehicle that looks the business, purposeful and practical with obvious off-road capability. It has a modern shape with a distinctive family likeness echoing its Land Rover heritage but integrated in a design appropriate to the 1990s.*

*On the inside it has a design which took shape for the way people live now. That's to say it takes account of the sports people pursue, the leisure activities they undertake and the ever-widening range of day-to-day uses to which people need to put their vehicles. It has wide-angle visibility for maximum safety and a no-compromise approach to passenger comfort and interior appearance. You'll find the Discovery has a superbly integrated design with coordinated theming of shape, colour and texture.*

*It isn't just our thought which has gone into the vehicle. We took the trouble to ask customers in the UK and throughout Europe what they were looking for in a vehicle of this type. The whole design concept of Discovery has been based around their wishes and needs.*

*In all aspects of performance – from speed to maximum torque, from acceleration to its all-important towing capability – Discovery is best or class-competitive. Two engines will be available. There's the familiar 3.5-litre V8 petrol engine and a brand-new 2.5-litre four-cylinder turbocharged and intercooled direct-injection diesel engine. The new engine is called the 200 Tdi.*

*Discovery has been tested and proven in the most arduous climates and in on- and off-road conditions around the world to satisfy our stringent engineering objectives for fuel efficiency,* reliability and low cost of ownership. Discovery has already covered in excess of 2 million miles and come through with flying colours.

Finally in terms of value for money this class-leading new vehicle will be on sale tomorrow at prices starting at £15,750 for the V8 or the 200 Tdi. In addition we have a number of option packs and we've developed a full range of fifty accessories and options from headlamp guards to ski holders together with a superb selection of leisure-wear products.

In a few moments the car will be revealed to you here but to really experience it we hope that later you book a test drive with my sales staff. Before we unveil Discovery please join us in watching a short video presentation of the car going through its paces and our stunning new TV and cinema commercial. After that the formal part of tonight's celebration will be concluded with a brief address from [name of Land Rover manager]. Play video.

The Land Rover representative's speech script was as follows.

Good evening. My name is [name] and I am the [position] at Land Rover. Tonight is a very special occasion – not just here but at every one of the Land Rover dealerships throughout the country. Every one of them will be holding a privileged preview of Discovery before it goes on sale tomorrow.

The entire management of Land Rover, including Sir Graham Day, are fully supporting our dealer network on this most important night and in the days to come.

Discovery is our first all-new car for twenty years and the launch of a new Land Rover is unlike that for any other vehicle. We are a manufacturer that sets standards – we don't aim simply to match the standards of others. When a Land Rover launches it is an event in the true sense of the word.

Over forty years ago a Rover engineer called Maurice Wilks

*dreamt of a vehicle able to do anything. In 1947 his dream became a reality with the very first Land Rover. Since that time, that rough, tough, far-from-elegant vehicle has earned its reputation as the world's best four-wheel-drive workhorse. Today that vehicle is four-wheel-drive in the minds of literally millions of people – so much so that it has defined the sector. Say the name Land Rover and immediately you have a picture of a certain type of motoring.*

*In 1970 we did the same again with the launch of Range Rover – a vehicle which represents luxury four-wheel driving. Again an innovation. Again it's a vehicle which has consistently set the standard in its field.*

*Now we launch Discovery – a new vehicle in a new sector of the market, very different to the two sectors Land Rover created in the first place. It's a vehicle which draws on a long Land Rover tradition and proven pedigree, but at the same time it's a vehicle designed specifically for the way people like yourselves live today. It's a true multipurpose vehicle for lives which are becoming more varied and demanding, a vehicle which will be equally at home in the town and the country, off-road or on-, business or leisure.*

*We are aware that for us Discovery represents a challenge and an opportunity. The challenge has been to provide a vehicle which we are confident will set the standard in its group. The challenge has also been to provide a vehicle for many customers who may not have considered Land Rover before.*

*The opportunity is to make sure that both the vehicle and the service and support we aim to provide will not only meet but exceed your expectations. Already it has generated a huge amount of enthusiasm since the first ideas were put together back in 1986. At the factory during its design-and-development phase; during its gruelling worldwide testing programme; from the press during their extended launch and driving appraisal; now by our dealers and their staff, both here and around the country tonight.*

> *The time has come though to hand over to the people that really matter – you! It is your appreciation and loyalty which have made the success of Land Rover in the past. I'm delighted that you are here tonight to share in our own excitement, to witness the arrival of the latest standard-bearer from Land Rover. Ladies and gentlemen, the new Discovery. [Dealer principal reveals Discovery to the audience].*

It's not often that one can look back to what happened over thirty years ago and recall in detail the words used and the contribution made by the dealership management and the Land Rover company representatives in launching a most important vehicle in Land Rover's history.

I have referred to the fact all the UK Land Rover main dealers that launched the new Land Rover model did so with a consistent presentation speech from the dealer principal and a similarly consistent speech from the Land Rover representative.

What follows for posterity from my original launch schedule is a full listing of all the Land Rover dealers, the dealer principals and the Land Rover representatives that were involved in what I consider to be a unique new vehicle launch.

### Discovery Launch 15th November 1989 (*16th November 1989)

| Land Rover Dealer | Dealer Principal | Land Rover Representative |
| --- | --- | --- |
| Aberdeen Land Rover Centre | Donald McHardie | Steve Yardley |
| Appleyard, Ayr | Alec Baird | Barbara Soar |
| Appleyard, Edinburgh | Peter Martin | Allen Murdoch |
| Corrie, Dumfries | Tony Thornton | Bill Wylie |
| County Motors, Carlisle | Mike Sherrard | Graham Silvers |
| Croall Bryson, Kelso | Robert Croall | Colin Andrews |
| Donnelly Bros, Dungannon | Terence Donnelly | Charlie Sneddon |

| Land Rover Dealer | Dealer Principal | Land Rover Representative |
|---|---|---|
| Dutton Forshaw, Sunderland | Barry Davis | Steve Whiteley |
| Gaulds, Glasgow | Stanley Gauld | Brian Purves |
| Hadwin, Torver | Eric Hadwin | David Crawley |
| Charles Hurst, Belfast | Ken Surgenor | Charlie Sneddon |
| Macrae & Dick, Inverness | Jim Chalmers | George Hassall |
| Minories, Middlesborough | Brian Foskett | Mike Gould |
| Morrisons, Stirling | Logan Morrison | John Lumsden |
| North East Motors, Newcastle | Alec Colvin-Smith | David Fulker |
| Frank Ogg, Aberlour | Frank Ogg Sr | Barry Tanser |
| Cowies, Perth | Graeme Moyes | Jim Orr |
| SMT, Greenock | Russell Dunn | Andy Bruce |
| JBW Smith, Cupar | Ron Smith | Dennis Rees |
| Taggarts, Motherwell* | Joe Nimmo, John Gillespie | Dennis Rees |
| T. P. Topping, Enniskillen | Peter Little | John Hamilton |
| Central Garage, Malton | Noel Calvert | Peter Armel |
| Central Garage, Driffield | Richard Briggs | Peter Armel |
| Armstrong Massey, York | David Turner, N. Brown | Steve Broadstock |
| Ripon Land Rover Centre | Geoff Brown | Richard Fox |
| Appleyard, Leeds | M. Williamson, C. Welch R. Hirst | Sir Graham Day, John Sparrow |
| Hatfields, Sheffield | Andrew Jeffrey | Roland Maturi |
| Farnell, Bradford | Ian and David Farnell | Tim Goldthorp |
| Rocar, Huddersfield | Nick Szkiler | Alan Guest |
| Ribblesdale Motors, Settle | Roy Melsome | Adrian Stewart |
| Southern Bros, Blackburn | Bill Rostron | James Batchelor |
| Lex, Bury | J. Chappell | David Marsden |
| Hollingdrake, Stockport | Phil Darragh | Todd Sharvell |

| Land Rover Dealer | Dealer Principal | Land Rover Representative |
| --- | --- | --- |
| Dutton Forshaw, Preston | B. Taylor | Steve Westwood |
| Currie Motors, Liverpool | M. Harvey | Martin Vine |
| Tannery Farm Garage, St Helens | Harry Jacoby | John Smith |
| Drabble & Allen, Knutsford | S. Hall | Gary Groves |
| James Edwards, Chester | Tom Booth | Bill Travis |
| Conwy Land Rover Centre, Llandudno | Peter Hewson | Neil Johnson |
| Mylchreest, Isle of Man | David Mylchreest | Ian McKay |
| Yarnolds, Stratford-upon-Avon | John Yarnold | Terry Haswell |
| Boston Tractors, Boston | Bryan Dobbs | Ken Plant |
| Colliers, Birmingham | Alan Clark, Phil Nye | Phil Breckon |
| County Motors, Lincoln | Alan Eades | John Evans |
| Duckworth, Market Rasen | Martin Duckworth | Sam Cinnamond |
| Droitwich Garage, Droitwich* | David Street | Mike Arthur |
| Evans Halshaw, Solihull | Jeremy Snowdon | Paul Kirk |
| Evans Halshaw, Stoke | Geoff Lovatt | Dick Elsy |
| Hartwells, Wellingborough | Robin Hall | John Baumber |
| Hartwells, Banbury | Bruce Houliston | John Hambleton |
| Henlys, Coventry | Peter Bruton | Barry Murray |
| Wadham Kenning, Ripley | Bernard Asher | John Stephenson |
| Kenning, Shrewsbury | David Wilks, Omar Kador | Bill Weald |
| Lex, Stourbridge | Mark Hickman, Jeremy Morgan | Harry Dunlevy |
| Mann Egerton, King's Lynn | Brian Fawcett | Adrian Morris |
| Mann Egerton, Norwich | Earl Tumilty | Bill Thomas |

| Land Rover Dealer | Dealer Principal | Land Rover Representative |
|---|---|---|
| Marshall, Bedford | Keith Howlett | Ian Upton |
| Marshall, Cambridge | Maurice Prove | Mike James-Moore |
| Marshall, Peterborough | Gerald Gough | Roger Panton |
| Shukers, Ludlow | Garth Joscelyne, Stuart Corrie | Colin Vaughan |
| Stafford Land Rover Centre | Dennis Bunning | Kelvin Alexander |
| Sturgess, Leicester | Robin Sturgess | John Cunnane |
| TMS, Melton Mowbray | Peter O'Connor | Alex Stephenson |
| Trinity Motors, Hinckley | Bob Woodward, John King | Robin Cardwell |
| Wiggs of Barnby, Beccles | Derek Wilkins | John Holford |
| Wolverhampton Land Rover Centre | Terry Wooldridge | John Rutherford |
| Merlin, Nottingham | Steve Hopewell | Chris Langton |
| Minden, Bury St Edmunds | Andrew Spencer | Stephen Schlemmer |
| Barretts, Canterbury | Geoffrey Barrett | Harold Kermode |
| Beadles, Dartford | Peter Liddle | Shaun Jefferies |
| Caffyns, Lewes | Eddie Ansell | Rob Pugh |
| Dunham & Haines, Luton | John Dunham | Mike Cheeseman |
| Dutton Forshaw, Aylesbury | C. Wyatt | Dave Bestwick |
| Dutton Forshaw, Maidstone | Mike Newman, A. Clarke | Colin Green |
| Follett, City of London | Graham Kimberley | John Russell |
| Guy Salmon, Ewell | Frances Thomson | Bob Francis |
| Harwoods, Pulborough | Les Sparkes | Alan Simpson |
| Hexagon, London | Roger Perks | Don Groves |
| Hunt Grange, Lamberhurst | Tony Ryan | Keith Lane |
| Ian Allan, Virginia Water | David Larmuth | Ken James |
| Lex, Maidenhead | Colin Harber | David Wilkins |
| Loxleys, Bromley | Fred Archer | Colin Kitching |

| Land Rover Dealer | Dealer Principal | Land Rover Representative |
|---|---|---|
| Mann Egerton, Bishops Stortford | R. Mills | Phil Payne |
| Mann Egerton, St Albans | Keith Woods | John Cooper |
| Mann Egerton, Colchester | Clive Chapman | Mike Hodge |
| Mann Egerton, Ipswich | Barry Sargent | Mike Stevens |
| H. R. Owen, Greenford | T. Brown | George Adams |
| H. R. Owen, South Kensington | Richard Crosthwaite | George Adams |
| Skinners, Hastings | Ted Wilmoth | Les Geary |
| SMAC, Southend | T. Wrigley | Alan Stedall |
| Southern Counties, Crawley | Gus Donald | Chris Woodwark |
| Stratstone, Mayfair | James Smillie | Ken James |
| Testers, Edenbridge | Paul Kentish | Peter Mackie |
| Wadham Stringer, Guildford | John Phillips | Brian Anderson |
| Woburn Abbey Garage | Rex Holton | Bob Hester |
| SMAC, Chelmsford* | Brian Keene | Alan Stedall |
| Bristol Motor Company | Brian Kendall | Andy Green |
| County Garage, Barnstaple | Patrick Squire | Bill Warburton |
| Exeter Garage | Peter Russell | Terry Donovan |
| M. J. Fews, Wotton-under-Edge | Mike Fews | Lou Lees |
| Fletchers, Swansea | David Dossett | Brian Price |
| Greens, Haverfordwest | Malcolm Green | Iorrie Williams |
| Hartwells, Bath | Chris Daubney | David Nicol |
| Hartwells, Oxford | John Pennie | Mike Donovan |
| Hereford Land Rover Centre | Nick Jenkin | Bob Stanton |
| Julians Land Rover, Cardiff | N. Watkins | Vic Shayler |

| Land Rover Dealer | Dealer Principal | Land Rover Representative |
|---|---|---|
| Julians, Reading | David Lamb | Alan Edis |
| Lex, Cheltenham | Barry Oakhill | Elaine Timmins |
| Lex, Newport* | J. Roberts | John Watkiss |
| J. V. Like, Hay-on-Wye | Mike Like | Tony Davies |
| Lloyd Motors, Aberystwyth | Evan Lloyd | Andy Gibbons |
| Olds, Dorchester | Peter Old | Barrie Wood |
| Ottons, Salisbury | Mike Otton | John Bragg |
| Riders, Falmouth | Ralph Sweet | Ron Parr |
| Ruette Braye, Guernsey | Jeff Kitts | Chris Batiste |
| St Helier Garage, Jersey | Geoff Habin | Neal Jauncey |
| T. H. White, Wootton Bassett | John Baker | David Carpenter |
| Taunton Garages | David Carr, Chris Spaett | Peter Wyhinny |
| Vincents, Yeovil | David Vincent | Mike Slattery |
| Wadham Stringer, Plymouth | Mike Dion | Les Pinkham |
| Wadham Stringer, Southampton | R. Higham | Tim Shilvock |
| Wadham Stringer, Waterlooville | J. Goodacre | Chris Mark |
| Webbers, Basingstoke | Brian Horsnell | Tom Whittaker |
| Westover Motors, Poole | Peter Wood | Bob Allsopp |

According to my handwritten notes additional members of company management provided support to the Land Rover representatives and the dealers, so for completeness of the company's involvement and for their contribution to the Discovery launch I am including their names and the Land Rover dealers they supported. I cannot guarantee the dealers they visited are totally accurate as I know some last-minute changes took place in terms of who went where. The following is from my notes in no particular order.

John Walsh at Taunton Garages; Andy Gore at H. R. Owen, London; Will Hope at Hexagon, London; Bob Burns at Colliers, Birmingham; Roger Hewitt at Henlys, Coventry; Martin James at Hartwells, Oxford; Helen Hughes at Lex, Stourbridge; Steve Biddle at Evans Halshaw, Solihull; Roger Conway at Appleyard, Edinburgh; Tony Easter at Sturgess, Leicester; Chris Scaife at Rocar, Huddersfield; Peter Gale at Ottons, Salisbury; Martin Luxton at Wadham Kenning, Ripley; Brian Turner at Follett, City of London; Les Goodwin at Minden, Bury St Edmunds; Peter Knight at Julians, Reading; Robert Furio at Merlin, Nottingham; Ian Duvoisin at Stafford Land Rover; A. Bourne at Droitwich Garage; C. Danks at Hollingdrake, Stockport; G. Johnson at James Edwards, Chester; B. McDonald at Lex, Bury; S. Rose at Wolverhampton Land Rover; R. Smith at Evans Halshaw, Stoke-on-Trent; G. Taylor at Merlin, Nottingham; and T. Worthington at Hatfields, Sheffield. Unfortunately there are a number of people in the above listings whose first names were not recorded.

16th November 1989 was the public launch date for the Discovery in the UK and a few dealers decided to hold their event that day. It was clear from the reports and feedback I received that the UK dealer events launching the Discovery had been a great success.

On 23rd November, having received, reviewed and analysed the individual reports from the Land Rover representatives who attended the dealer launch events, I produced and sent a Discovery launch report to Chris Woodwark, Land Rover Commercial Director, John Russell, Land Rover Sales and Marketing Director, John Cunnane, UK Sales Director, and Brian Purves, Finance Director. I am reproducing this report word-for-word as follows.

*Discovery UK Launch Report*

*1 Dealer Launch Activity*
*On 15th November 1989 all 127 main dealers held VIP Preview*

*presentations of Discovery – a few in fact were held on 16th November. On this occasion all dealer principals introduced the new Discovery and all dealerships were supported by Land Rover senior management, demonstrating Land Rover's commitment to the Discovery and to our dealers and customers.*

*Conservatively, the Discovery was launched to some 25,000 potential customers and I estimated from launch plans that the UK Land Rover dealers collectively had invested a total of £1/2 million in producing prestigious and professional launches.*

*On the night a number of brand-new dealerships were also unveiled, demonstrating the commitment of the network to the Land Rover franchise. Notable among these were new facilities declared open in Leeds (Sir G. Day), Crawley (C. J. S. Woodwark), London (J. K. Russell) Llandudno (N. A. Johnson) and Pulborough (A. K. Simpson).*

*Most dealers had a good cross-section of potential conquest customers, and while there will be some unavoidable trading-down from Range Rover the main dealer concern is the effect of Discovery on used two- to three-year-old Range Rover sales.*

*2 Customer Reaction*
*From the Land Rover management responses the overall reaction is excellent. In particular Discovery was praised on its competitive price, styling, interior space, 'car-like' comfort. There was a notably good positive reaction from the females. The main areas of concern raised were absence of a five-door derivative, automatic (southern England) and vehicle supply.*

*3 Projected Sales Forecast/Allocation/ Sold Orders*
*The 1990 UK dealer objective is 4000 Discoveries while the current 1990 sales forecast is 5000 (three- and five-door). It is already evident the average dealer has sold its initial launch allocation (total*

*UK volume of 1000 units) and is quoting between three and six months' delivery. This would imply dealers have sold orders of additional 1000–2000 vehicles. Obviously we expect there will be a degree of order duplication as customers are 'ringing around' to place orders. However, it would seem we have a high level of dealer discipline and support to the marketing of Discovery.*

*While price is considered very competitive, and the Discovery is considered excellent value for money, dealers are ordering high-specification derivatives, i.e. vehicles which will retail from £18,000–19,000 before they add on accessories (an area where demand outstripped supply on launch).*

*It is clearly important we monitor the first few months of Discovery in order to establish meaningful sales forecasts/allocations for the UK. However in order to ensure disciplines are maintained while recognizing demand levels, allocations to the UK need to be around 450 units per month from March giving a full 1990 total of around 5400 units.*

*At these levels the projected allocations for UK of 335 units (March), 210 units (April) and 390 units (May) will be inadequate by some 415 units over these three months. UK Sales requests an urgent review to ensure a more appropriate level of allocation to support the UK dealer investment in the Land Rover franchise and to satisfy demand.*

*J. A. Sparrow, 22nd November 1989, JAS/TEW/036/107*

To all the Company Directors and Managers who represented Land Rover at a dealer launch of Discovery I sent the following memo.

*It is evident from the response forms I have received that the launch of Discovery on 15th November was very successful.*

*All dealers have reported an extremely positive response to Discovery and the programme of formally launching Discovery with the direct involvement and support of Land Rover representatives was clearly beneficial to our dealers and customers. In addition I am sure your presence at Land Rover dealerships will have given you all a very useful and memorable experience, particularly those of you who do not normally come into direct contact with our customers and dealers.*

*The response forms containing specific details of customer likes/dislikes/reaction etc. will now be passed to Marketing for analysis and action. I would like to take this opportunity to thank you for your cooperation and support.'*

J. A. Sparrow, 29th November 1989, JAS/TEW/036/128

There were many positive articles in the press about the new Discovery. I will just repeat the words of the one that appeared in the *Today* newspaper of 30th November. Headed, 'Year-Long Delay on Road to Discovery', it continued as follows.

*Executives are being forced to wait a year to buy a sparkling new Land Rover which is leaving its Japanese rivals standing. Some are even offering to pay thousands over the £15,750 price tag to get their hands on the all-terrain Discovery. Dealers cannot keep up with demand for the scaled-down Range Rover which is out-gunning Japanese market leaders such as the Mitsubishi Shogun.*

*Production of the Discovery is being stepped up from 250 a week to 300 to cope with the rush. Output will be up to 390 in January less than three months after its launch. Production is on target to overtake the super successful Range Rover now being made at 640 a week.*

*'We have been absolutely overwhelmed by the demand' said Phillip Nye of Midland dealers Colliers. 'We have had a lot of people with Japanese vehicles wanting to switch. They even seem prepared to wait nearly a year to get their hands on one'. Mr Nye added 'Motorists have just been longing for someone to produce a British-built competitor to the Japanese off-road vehicles. The Discovery is certainly the best of British.'*

*Autocar* magazine, after comparing it with the Isuzu Trooper and Mitsubishi Shogun, concluded Discovery was:

> *Faster, more economical, better riding and with the extra traction and balance of permanent 4wd had the measure of its rivals. The cleverly designed, well-executed interior is way ahead of the opposition and it has a clear advantage should anyone actually venture off-road ... Where Land Rover will struggle initially is in trying to sell the Discovery to people who have grown used to five doors – but its other qualities more than make up for a slight struggle to get into the rear.*

Within Land Rover it was the successful launch of a new third model in the range that we had all been planning, and which would now give the dealers in the UK network the justification for the investment and commitment to the Land Rover franchise they had made. There were many dealers a few years earlier who had decided not to invest in the Land Rover franchise and some of these would now come to regret it.

In 1990, the first significant model improvements were introduced. The V8 engine received the Electronic Fuel injection (EFi) system it should have enjoyed from the beginning, and the five-door model, using Range Rover doors, was introduced. Discovery had quickly established itself as the leader in its sector of the market, and we were very pleased to see that there was almost no adverse impact on new sales of the Range Rover. In practice there were opportunities to move the Range Rover further up-market in the luxury-car sector.

Towards the end of 1989 we were forecasting sales of 450 new Discovery models a month in early 1990, admittedly before the five-door model was introduced. In some sort of comparison there were reportedly 5,577 sales of the new Discovery Sport alone in April 2015, albeit this was a model that replaced the Land Rover Freelander. In 2015 the base-model Discovery Sport started at £30,695, while the cheapest Range Rover (the Evoque) started at £30,200. The base-model Discovery price at launch in 1989 was £15,750.

Some 56,200 Land Rovers were registered in UK in the full year 2014, representing 2.27% of the total new car market. I don't think Land Rover sold that many new vehicles in the whole world back in 1989. The UK was the largest Land Rover market at that time and there was no market opportunity for Land Rover vehicles in major markets such as China, India and Russia as there is today – and the term 'emerging markets' hadn't yet emerged!

In the USA between 1974 and 1987, Land Rover vehicles were only sold in relatively small numbers through the grey market. Land Rover began selling the Range Rover officially in the USA on 16th March 1987. From that time until 1993, all marketing was in the name of Range Rover, because it was the only model officially offered in the American market. In 1993, with the arrival of the Land Rover Defender 110 and the imminent arrival of the Land Rover Discovery, the company's sales in the USA were made under the name 'Land Rover North America'.

The 1994 model-year Discovery marked the first year that the Discovery was sold in the United States. Airbags were incorporated into the design of the 1994 model to meet the requirements of US motor vehicle regulations, though they were not fitted as standard in all markets. All North American-specification models were fitted with the 3.9-litre V8 from the Range Rover SE model, and later models saw a change to the 4.0-litre version of the engine.

# Chapter 12
# Land Rover Ltd End 1989–1990

**ON FRIDAY 24TH NOVEMBER** it was officially announced that John Cunnane would be leaving Land Rover. However, it was not until 12.30 p.m. on Monday, 27th November that the formal announcement was made to all Land Rover staff that I was to be Land Rover's new Director UK Sales, and that this would take effect from 1st December 1989.

I had a little time to contemplate the challenge of the Director UK Sales role before the announcement was made. I had to consider what my new position would mean in terms of my key activities, and to contemplate what changes I might want to make to strengthen the UK sales organization. I was aware I would need to delegate much more than I had in the past.

John Russell reminded me of this and I remembered that John Cunnane had told me, when I became the UK Sales Operations Manager, my job was to manage while his job as Director was to direct me. I now had to direct, not manage.

There were several important organization decisions to make. First, who should be my successor, and second, what, if any, changes I should make to the existing sales structure. In the two days after the internal announcement of my new role I had meetings with John Russell and John Cooper, Land Rover's HR Manager, not only to express my initial views but also to get theirs. I had some thoughts about my successor from within the existing UK Sales team. However, it was suggested that the company wished to broaden the sales experience of Rob Pugh, who at that time was the UK Parts Sales Manager within Land Rover Parts Ltd, and as I had no objections, his appointment was agreed.

With regard to the Regional Sales Manager team, it was agreed some new impetus to the team was desirable in order to drive through the next stages of growing the UK Sales business. I believed that it was necessary to strengthen the regional sales team with a few potential leaders of the future and one person I had in mind for this was Andy Bruce. I had been given the opportunity and responsibility at a relatively young age to work an area with UK dealers and I decided Andy, a Scot living in Glasgow, should take over the Scottish region.

We also decided to bring Steve Westwood and Paul Evans into the Regional Sales team covering the Northern and the Western region respectively. Tim Goldthorp moved from the Northern region to a new role as UK Sales Support Manager, combining the previous sales and fleet sales administration functions. Tim would be supported by John Walsh, Sales Administration Manager, Bob Burns, Jim Cooke, Albert Charles and Mark Valente. David Carpenter moved from the Western to the Midlands region, and Bob Hester continued his role as Regional Sales Manager for the Southern region.

We created a new Major Accounts function headed by John Lloyd who moved from his previous Technical Service Manager position. In this role John was not only responsible for sales to all Direct Account Fleet Operators but also for the development of sales for Land Rover's Special Vehicle Operations (SVO) department. He would be ably supported by Ron Parr and John Smith, two very experienced and knowledgeable executives.

Barry Tanser transferred from Marketing to be UK Franchise Development Manager. In practice his responsibilities were largely unchanged but he would now devote his time fully to UK facility development programmes, the implementation of Land Rover corporate identity, premises development projects and franchise décor schemes.

At the same time that John Cunnane left the company, so did his secretary, Win Parr, the wife of Ron Parr. Tracey Wilmot, my previous secretary, would now support Rob Pugh. Fortunately I was able to take on Kath Duval as my new secretary, a lady who had worked previously for John Sewell, another Land Rover Director. Kath would prove to be excellent in supporting me in my new role.

In considering and carrying out these structural and personnel changes in the Land Rover UK Sales function I was aware and reminded of an analogy from football about the importance and benefits of strengthening an already very good team. Liverpool in the late 1970s and 1980s was one of the most successful football teams in Europe, but its team management continually added a few new younger talented players into the team, with a view to improving the quality and competence of the team and keeping them ahead of the competition and maintaining levels of success.

With the agreement of John Russell and John Cooper on the full range of proposed personnel changes I decided I would speak to everyone involved in the department individually as soon as possible about my plans and their own situation. This would all be done before the Christmas holiday, which was only a few weeks away. In reality many members of the team would continue in their existing roles, but for the few who would be subject to change it was important to tell them at the earliest opportunity. I made a note in my diary to have individual briefing meetings with everyone affected by change and as many as I could throughout the day on 14th December and onto the following day as necessary.

On 1st December John Russell circulated a usual organization and appointments communication within the Rover Group confirming my appointment as Director UK Sales. Other appointments confirmed or announced were Roger Charlton as Marketing Communications Director joining from Rover Cars, Alan Edis as Marketing Plans Director, Roland Maturi as SVO Director, Chris Langton as Service Director, Alan Guest as Export Sales Director and George Adams as Government and Military Sales Director.

On 14th December accompanied by John Cooper, I held individual briefing meetings with twelve members of the UK Sales team, explaining my plans and confirming their personal situation. As everyone knew of their meeting time with me, I was aware many would be very anxious about it and what I might have to say. And so it turned out that a number were more worried than they needed to be and were very relieved when I confirmed they would have a continuing role in UK Sales.

There was not a better way to communicate what I had to say to everyone than to do it individually, face-to-face. These meetings included those where I had to confirm the difficult decisions to replace a few long-standing members of UK Sales. The following day I spoke to the remaining other members of the UK Sales team either face-to-face or in a few cases over the phone about their individual position.

During the last few months of 1989 I was involved in an interesting franchising project that I had initiated in Coventry. This city had been designated one in which we were looking to have an exclusive Land Rover dealer. At the time Henlys was the Land Rover dealer in Coventry, operating this alongside Rover Cars. However, it was unable to submit proposals to meet Land Rover's requirements within the necessary timescale and so it was informed we would explore other options.

At the same time Tom Walkinshaw Racing (TWR) represented Jaguar Cars in Coventry from its premises on the A45, very close to Browns Lane, where Jaguar had its Head Office and main production plant. On 21st December I had a meeting with Tom Walkinshaw, the owner of the Jaguar dealership, to see if the Land Rover franchise in Coventry would be of potential interest. I was pleased to get a positive response. He told me he had plans to relocate Jaguar to new premises, and in principle he would be willing to take on Land Rover once he had had relocated the Jaguar business. This duly happened and on 1st May 1990 the new exclusive Land Rover dealership in Coventry called Shires opened.

Prior to the meeting with Tom, I discussed with Chris Woodwark and John Russell the implications of appointing TWR since the removal of the Land Rover franchise from Henlys would not be supported by Henlys and Rover Cars management. And it turned out I received strong opposition from Rover Cars on the basis the loss of the Land Rover franchise would seriously undermine the viability of the Henlys and the Rover Cars business. This was not a unique response from Rover Cars top management to Land Rover's franchising strategy, but Canley in Coventry was at that time the Rover Cars Head Office location, so the proposal to terminate Henlys and appoint TWR was perhaps seen as hitting Rover Cars in its home location. And so ended what was for me a very memorable 1989.

The year 1990 would be no less busy for me. On 11th January I sent out a letter to all the UK dealers confirming full details of the new Land Rover UK Sales organization. This was followed later in the month by a series of five Regional Dealer Meetings to present our 1990 business objectives and operational plans.

At the end of 1989 I was asked to join a new project team, Rover Group Image Breakthrough. The aim of the project was to look at all areas of the business and to identify specific actions that would improve Rover Group's image. The team comprised Land Rover and Rover Cars senior management and we seemed to meet every few weeks at the company's offices in Canley, Cowley or Solihull. The meetings continued in 1990 every few weeks through to the end of the year.

I can't recall any significant outcomes or initiatives, although we did spend much time reviewing the image of our vehicles, and discussing what could be done to improve that. I remember saying a driver of a car should be able to look in the rear-view mirror and immediately be able to identify the car behind by its distinctive front grille, as in the case of BMW or Mercedes, brands whose vehicles had distinctive family grilles on all their models. At the time Rover's cars just had a relatively small Rover badge on the front of the bonnet. I thought this was weak in comparison to the German brands. Subsequently the Rover 400, 600 and 800 did acquire a distinctive Rover grille so I suppose there was some benefit from the project.

In January 1990 I produced a 'UK Dealer Franchise Strategy

Status Report' for John Russell copied to Chris Woodwark. My report provided a detailed analysis of the shape and size of the Land Rover dealer network. Of the 129 main dealers, eighty-four operated from totally separate premises. Of the forty-five dealers whose facilities were shared with another franchise, thirty-four were shared with Rover Cars and six with Jaguar.

I commented that over the next two to three years there would probably be a need for some refinement in the network. I added that the need for higher levels of dedication and specialization may identify a need for 'new blood' of the sort successfully appointed in 1989, for example, Charles Follett and Hexagon in London. I concluded my report saying, 'The Franchise Operating Committee will meet quarterly to progress these issues.'

At the end of February 1990 I had a Directors' meeting at Gaydon to discuss the introduction of the Discovery five-door. This model would now address one of the main customer complaints at the initial launch. I cover more on that subject later.

During May of 1990 Annette and I were invited to a Scottish Motor Trade Association (SMTA) Rally in Scotland at the Gleneagles Hotel, Auchterarder. We drove there on the Friday, a journey of some six hours. An attraction for me was playing golf on the Gleneagles King's Course, which I regard as one of my favourite golf courses. It was a tough course to play, but enjoyable because of the fantastic scenery.

Annette reminded me of some words exchanged when she was driven around the off-road course by Andy Bruce. Someone came to speak to Andy through the vehicle's window and asked, 'Who is your lady friend?', to which Andy replied, 'The boss's wife.' Potential embarrassing situation avoided!

On 30th May I played in my first PGA (Professional Golf Association) pro-am tournament, the Dunhill British Masters, on the Duke's Course at Woburn. Our team comprised Brett Ogle, the Australian professional golfer, Rex Holton, Managing Director of Woburn Land Rover, Brian Allen, the actor Gareth Hunt and me.

Gareth Hunt was probably best known for appearing in the 1970s TV series *The New Avengers* alongside Patrick Macnee and Joanna Lumley. Gareth must have been appearing in one of Shakespeare's plays at the time, because on the tee he frequently said to one of us, 'After you, my liege.' Gareth certainly made the round of golf friendly and enjoyable even though we were in a serious competition.

This was to be the first of seven PGA pro-am golf tournaments in which I played during the period 1990 to 1996. There were to be three more at Woburn. I have photographs of all seven pro-am golf teams in which I played and these photos are grouped together on one of my bedroom walls. They are a great, almost daily reminder to me of taking part in pro-am golf tournaments and how fortunate I was to do so.

On 15th June 1990 I visited the BBC at its Pebble Mill Studios in Edgbaston, Birmingham. I was invited to talk about the Range Rover, as it was now twenty years since it had been launched in 1970. I discussed the significant development of the vehicle in that time as well as its successful sales growth. I was interviewed live on BBC Radio by Charlie Neil, who was better known later as a weather presenter on Central TV.

June was also the month when Land Rover had its usual presence with a large stand at the Royal Highland Show, near Edinburgh. However, in 1990 Land Rover Marketing arranged for a Land Rover Social Day for the Scottish dealers to take place at the Gleneagles Hotel the day before the Royal Highland Show opened. Once again every dealer who played golf signed up to play the King's Course. At the end of the week we watched the Jackie Stewart Celebrity Shoot at Gleneagles. The teams were made up of members of the Royal Family, as well as various European dukes and celebrities.

During 1990 a number of new exclusive Land Rover dealerships were opened in the UK. It was now normal for Captain Mark Phillips and me to present at these fairly regular premises openings. We were almost a double act. I would speak about the importance of the Land Rover dealerships in meeting the ever-increasing demands and requirements of customers, the importance of our customers, as well as the work we were doing to develop and introduce new Land Rover vehicles. Mark spoke about his work with the Range Rover Team and with his eventing activities. Mark's presence at these events naturally drew a full house and it was not surprising that the people attending these dealer openings were more interested in what he had to say than in my remarks.

As mentioned, after striking up a relationship with Captain Mark Phillips in 1980, Land Rover funded a programme of training sessions and bursaries for young riders, created and run by him. It was branded the 'Range Rover Team'. Two years later Land Rover became one of the original sponsors of the British Open Championship, which was first held and organized by Mark at Gatcombe Park in 1983.

At the end of June 1990 Mark and I attended the official opening of Matford Land Rover in Exeter. On this occasion I was fortunate to secure the services of Tony Cottrell, the company chauffeur, who drove me to Exeter and afterwards back home.

The following week I was back at work in Solihull and into a round of internal meetings that included a 1991 model-year pricing meeting with Land Rover Directors Alan Edis and Brian Purves, and a distribution efficiency meeting with Chris Woodwark.

August included a major Land Rover marketing event which made its debut in 1990 – Land Rover Cowes Week. A unique amphibious Discovery was built by Land Rover's Special Vehicles department to promote Land Rover's sponsorship of Cowes Week on the Isle of Wight. Annette and I went to Cowes for two days to assist with the hospitality Land Rover and its dealers were giving to their customers and invited guests. One afternoon we went out to watch the racing close-up and I could see the attraction of the yachting races at Cowes.

In my diary for the latter part of August there are many references and meetings regarding the forthcoming September conferences, which would in fact be a series of five 1991 model-year Regional Business Meetings to be held at the National Motorcycle Museum in Birmingham during the period 5th to 12th September. These conferences included the official launch of the Discovery five-door to the dealers, but by this time the dealers knew all about the vehicle. It had been impossible to keep it secret. There would be a dinner for the dealers and an overnight stay at the Hyatt Hotel in Birmingham.

The after-dinner entertainment each night was provided by John Lancelot Blades Percival, better known as Lance Percival. Lance was an English actor, comedian and singer, best known for his appearances in satirical comedy shows of the early 1960s such as *That Was The Week That Was*. He appeared in many comedy films in the 1960s and 1970s. Lance later became successful as an after-dinner speaker and in 1990 Land Rover Marketing hired him to do just that.

I was a little surprised to be asked by Lance to meet him before he was due to speak at the first dinner. However, he wanted to get the names of several members of the Land Rover top management and a few of my managers, and explained that these would be built into his after-dinner speech. It became clear that he had the overall format and wording in terms of stories and anecdotes already fixed and just needed to use Land Rover management names to personalize his speech for the Land Rover dealer audience.

Lance was also clever in his use of words. So rather than referring directly to the launch of the Discovery five-door, he had created a story about 'Dis lovely live wh***'. Definitely not politically correct now but he got away with it in 1990. His anecdotes included a fictitious one about him dropping in to see Annette and me at our home in Knowle for lunch. He had asked me if there were some of my managers who would not take offence at having their character attacked, if not assassinated, during his speech.

One of Lance's short anecdotes related to one of my Regional Sales Managers having to go to the dentist. The punchline was that 'he had gone there to have some wisdom teeth put in'. Lance was able to deliver his after-dinner speech without any notes, which was very impressive to see. He had memorized each of the names within the relevant anecdote and the dealers greatly enjoyed hearing Lance making some fun of the Land Rover management. I suppose you could call him a satirical after-dinner speaker.

Soon after I received a phone call from George Simpson's office. It was not about work, but would I captain a second Land Rover team in the Suntory World Match Play pro-am golf tournament to be held at the Wentworth Golf Club in Surrey on 19th September? George would captain the A-team. My answer was a simple yes. I was told Mr Miyake, Managing Director of Honda of the UK Manufacturing Ltd and Alan Tipping of the Ministry of Defence would be in the team. I just needed to find one other player to make up the team and so I contacted and received a positive answer from David Piggott of Charles Follett, a recently appointed Land Rover dealer in London.

I knew the West Course at Wentworth was a very difficult course and as I had no time to practise this event would turn out to be very nerve-wracking. During the 1980s, the Suntory World Match Play golf tournament was an event all the big names in golf played. Just twelve of the world's best players were invited and taking part in the 1990 tournament were Seve Ballesteros, Nick Faldo, Hale Irwin, Bernhard Langer, Mark McNulty, Greg Norman and Ian Woosnam. I mention Mark McNulty of Zimbabwe because I had his caddy for the pro-am event.

On the morning of the pro-am it was raining but not too hard. When it came to the time when we had to go to the tee I noticed there was a fairly big crowd surrounding it, which I had never encountered before. After Hale Irwin of the USA and his team had teed off it was our turn. This is when the tension really started.

I was announced by the starter when it was my turn to tee off. I walked over slowly and put a golf ball on the tee. I noticed several stewards hold 'Quiet' signs in the air. There was total silence from the crowd around me but I could hear a slight pitter patter of the rain on me and I was too aware of this. I eventually hit the ball. It was not a great tee shot but not the worst either. Walking off the first tee it also occurred to me that the people making up the big crowd had paid good money to watch the pro-am.

On this pro-am we were fortunately playing to a Texas Scramble format. This means each player in a team tees off on each hole and the team leader, invariably the professional, decides which shot was best. The other players then pick up their ball and play their second shot from the best-shot position. The procedure is repeated until the hole is finished with putting.

It was probably not until the third hole that my anxiety disappeared and I was able to try and enjoy this unique golfing experience. The Texas Scramble format removed the pressure of my having to hit long drives off the tee and long iron shots from the fairway. I remember I hit some good putts, which was my main strength.

I almost holed a very long putt on the 18th green and afterwards we realized holing my putt would have given us third place. Unfortunately no one else in the team could hole the putt and we finished fifth. The photograph of my team on one of my bedroom walls in Land Rover sweaters is an almost daily reminder of my first and only round of golf at one of the UK's most famous golf courses.

A member of George Simpson's team at Wentworth that day was Geoff Dale, Managing Director of the Evans Halshaw Motor Group based in Solihull. Geoff was someone I knew well since his company held the Land Rover franchise in the town, the home of Land Rover. At the end of the pro-am we had a conversation and I mentioned the fact I did not belong to a golf club and now I was playing in these pro-am tournaments I probably should do so to improve the standard of my play. Geoff suggested I should apply to join Ladbrook Park Golf Club, at Tanworth-in-Arden near Solihull, where he was a member. Geoff kindly said he would support my application.

The 1990 Motor Show was held at the National Exhibition Centre (NEC) in Birmingham from 19th to 30th September. It was at this show that Land Rover announced a limited-edition two-door vehicle called the

Range Rover CSK. I will write about my involvement in that vehicle because it was an interesting tactical product introduction. I say tactical because to the best of my knowledge it wasn't in the long-term product plan.

My involvement came one day when I was informed by my Director superiors that there was a proposal to build a special two-door Range Rover for the UK market. My initial reaction was that it wasn't serious, but I quickly realized it was a serious proposition. My business experience told me this was not something I wanted for the UK. Some background details may help explain my concerns.

The Range Rover was introduced in 1970 but only with two doors. Its specification was fairly basic and utilitarian, with the ability for the interior if muddy to be hosed out and cleaned with water. As the Range Rovers only had two doors, access to the rear seats was rather awkward and the two doors were also very large and heavy, with old-fashioned quarter-light windows. The introduction of a four-door Range Rover model was the first significant body change but this did not come until 1981. The four-door Range Rover was received well by customers and its popularity was such that the two-door model, with little or no demand, was discontinued in the UK in 1984.

Six years later in 1990 I was being told we were reintroducing a discontinued two-door Range Rover. The real reason soon became apparent. Land Rover had an availability of some 200 two-door Range Rover body shells that had no obvious home or market and so in considering possible options for their sale the idea of a special-edition two-door for the UK was conceived. It was thought that a UK dealer network of some 130 main dealers would have no problem taking, handling and selling 200 two-door Range Rover models. Senior marketing management decided these vehicles would be packaged as a sporty limited edition in black paint with each car having an individually numbered plaque on the radio panel, designed to confirm its exclusivity.

I was not keen on accepting these vehicles as I thought they would not be easy to sell. I did not share the optimism of the Marketing department on this venture. So I decided to sound out the dealer council members on this proposal and if this was to happen I wanted their support. When I phoned them many thought it was a retrograde step. However, I had to tell them that a two-door model was very likely to happen and that in the near future we would discuss the subject further including pricing and other relevant matters at the next dealer council meeting.

Prior to the dealer council meeting I had conversations with Chris Woodwark and John Russell to give them the views of the dealer council, and also to discuss pricing and the vehicle's name. The name Range Rover CSK was proposed, with the CSK representing the initials of Charles Spencer King, who designed the original two-door Range Rover. At the dealer council meeting I remember we asked the members for suggestions on names with the proviso a name should reflect something sporty and black. One dealer council member made a suggestion. However, the name CSK was then communicated and agreed.

We confirmed the car would be priced at £28,995 for the manual gearbox and £30,319 for the automatic version. Land Rover Marketing bought 200 number plates from H1CSK through to H200CSK to add exclusivity to this limited edition. The original owners of the CSK would receive a special sales brochure and a Range Rover CSK Owner's Pack, containing a black wooden box with an etched metal plate, the sales folder and a certificate of authenticity containing the original thank-you slip written and signed by Charles Spencer King.

When the Range Rover CSK was announced at the 1990 Motor Show, there was some scepticism from Range Rover enthusiasts. Most of them couldn't understand why anybody would want to buy a relatively obsolete two-door Range Rover in a market where only four-door models were being sold. One day during the Motor Show I was chauffeured to the NEC with Spen King, and he told me he had expressed some surprise when told about the CSK vehicle and said he personally had some doubts about the value of a two-door limited edition.

It should be noted that some two-door Range Rover models continued to be produced at that time, mainly for the French market, but not enough to have avoided the introduction of the CSK into the UK. In the end it was a costly exercise, but understandable when you have 200 unwanted body shells.

More new dealerships were being developed and on 3rd October 1990, Captain Mark Phillips and I travelled north to attend the official opening of the exclusive Land Rover dealership of Dutton Forshaw in Durham. A special guest was Jack Charlton, famous for being a member of England's 1966 World Cup-winning football team, and elder brother of Bobby Charlton, who was also in the 1966 team. At the time Jack was manager of the Republic of Ireland national football team and he had led them to their first ever World Cup in 1990, where they reached the quarter-finals.

Sometime in the summer of 1990 Annette raised the subject of moving house. We had been there since early 1983 and as we could afford to move to a better property we agreed we should start looking in the Knowle and Dorridge areas for a new home. Cala Homes were just starting to build new houses in Dorridge and any of these would be a big step up in specification and standard from our current property. Annette decided to progress with Cala.

To make the move attractive Cala Homes agreed to buy our house in part exchange, and at an acceptable price, for one of the first houses they would finish building. This would remove the pressure and estate agency cost of selling our house and would enable us to move straight into our new one. In October I placed a deposit on our new house, which was under construction. However, it would not be until 14th March 1991 that we finally moved out of our home in Knowle into 15 Morville Close, Dorridge.

Chapter 13

# Range Rover North America 1990

**IN THE LAST FEW MONTHS OF 1990** I became very involved in supporting Land Rover's business development in North America. Range Rover was launched in North America on 16th March 1987. The specification was a 178-bhp 3.9-litre fuel-injected alloy V8 engine, with a ZF four-speed automatic transmission. In 1988, its first full year, 3,427 Range Rovers were sold by the sixty-three dealers. In comparison, 6,175 Range Rovers were registered in the UK in 1988 by the 153 main dealers.

In 1989 Range Rover sales had grown to 4,822 in USA, through a dealer network of seventy outlets. 28,096 Range Rovers were sold worldwide (+16% over 1988). The USA had a population of 250 million in 1990 and with a car market around six times larger than that in the UK's 57 million population it would provide Land Rover with a major sales opportunity over the coming years.

In March 1990 I had my first meeting with Charlie Hughes, the President of Range Rover of North America (RRNA). We had achieved a great deal of success in reshaping the UK Land Rover dealer network from a low-volume franchise, mainly held alongside Rover Cars, into a largely exclusive and specialist 4x4 franchise.

Charlie now came to the UK fairly often and during this latest visit he wished to see some of these exclusive UK Land Rover dealers for himself. So on one day I took him to Lex, Stourbridge, Merlin, Nottingham and Yarnolds, Stratford-upon-Avon. While doing this I explained the strategy and action plans we had implemented, and the benefits we had derived in developing a network of dedicated and exclusive dealers. These dealers were 100% committed to Land Rover and as a result demonstrated this in all their dealings with customers. Not only did this help grow new vehicle sales but the dealers saw their efforts rewarded with greatly improved profitability. This would be the start of an increasing level of involvement and support I would give Charlie and his team over the next few years in the development of a network of Land Rover Centers (American spelling of Centre) in the USA.

It had been agreed that I would visit the USA to familiarize myself on the ground with the status of the Land Rover dealer network there and to pass on some of my knowledge and experience to the RRNA management and some of the more important dealers. A busy eight-day schedule would be my first business trip to the USA at the end of which I would attend and support a dealer-organized weekend off-road event for Range Rover customers in California.

On 11th November I flew to Washington Airport, where RRNA had arranged for me to be collected by a chauffeur and during the afternoon I was delighted to be given a guided car tour of some of Washington's main sights, including the White House. I was then taken the thirty miles or so eastwards to the Maryland Inn in Annapolis where I would stay for the first two nights of my trip. Annapolis, which is situated on Chesapeake Bay, is the capital of the state of Maryland, although it is much smaller than the city of Baltimore twenty miles to the north. Annapolis is also the home of the United States Naval Academy. Charlie Hughes lived in Annapolis, very conveniently placed just over twenty miles from the RRNA Offices at Lanham. On the Monday at Lanham I was introduced to most of Charlie's team including Roger Ball, the Marketing Vice President, who I already knew and who had joined RRNA from Land Rover at Solihull.

In addition to Charlie and Roger the executive team comprised William Baker, Vice President Corporate Communications; Joel Greer, Vice President Sales; James Lehmann, Jr., Vice President General Counsel; Joel Scharfer, Vice President Finance; Richard Hubert,

National Manager Parts; and David Schworm, General Manager Service. Most of the executives and management at RRNA had not long come from other motor manufacturers, recruited by Charlie into his team to meet the challenge of developing and making Land Rover a successful brand in the USA from an almost zero base.

During the day Charlie decided I should say a few words to everyone in the RRNA office about the purpose of my visit, as most would have wondered why I was there. So as soon as they were all gathered together in the main reception area I explained I was there to support the development of the RRNA business, using my knowledge and experience of having already helped develop and grow the Land Rover business in the UK. I explained the importance of creating a network of dedicated and exclusive dealers and that this should be the approach RRNA should adopt wherever possible. I added that with the much larger 4x4 vehicle potential in the USA than in the UK, and with additional Land Rover products to sell, the RRNA objective should be able to grow to be Land Rover's largest market.

On the morning of Tuesday, 13th November I flew with Joel Greer from Baltimore to New York. That morning we met with Roy Grace of Grace & Rothschild, RRNA's advertising agency.

Grace & Rothschild's innovative advertising at the time showed Range Rovers ploughing through rivers and climbing very steep slopes. One of their adverts showed a Range Rover coolly navigating its way across the side of a hill. Another advert showed a Range Rover driving through a stream. The copy read 'We brake for fish'.

I visited Range Rover Land, an exclusive dealer located at Glen Head on Long Island where the principal was Bryan Lazarus. At the time it was the largest Range Rover dealer in new sales in the USA. I remember the dealer was very well connected to the horse-owning community and that the sales manager was a lady who had been previously a Range Rover owner. She was also the salesperson with one of the highest sales of new Range Rovers in the USA at the time of my visit.

I was very impressed with the commitment and passion that Bryan and his team had for the Range Rover business and he would have an important dealer input into the development of Land Rover Centers. Later I took a flight from New York to Los Angeles, where I was met and picked up by Ken McGraw, who was responsible for RRNA's Western region.

On 16th November Ken McGraw took me to visit Hornburg, the Range Rover and Jaguar dealer located in Beverly Hills. I remember the visit very well because Bill Gaby, the dealer principal, showed me a book listing the names of his dealer's customers. A large number of the names were well-known Hollywood actors and actresses, but I recall Bill telling me that some of these names were very bad at paying for their vehicle's servicing and repairs. I wondered, did these wealthy people have more important things to do or were they hoping to avoid paying?

The next dealer we visited also made an impression on me for some of its interesting business activities. This dealer was Newport Imports, located on the coast at Newport Beach, and where the dealer principal was Lee West.

During a tour of the premises Lee outlined some details of his business. First I remember exchanging business cards with Lee and receiving the comment that mine wasn't a business card. He said mine was like most business cards, just a name, title and contact detail. His business card was double the size of a typical business card but folded so that there were four sides. On the front was a photo of the Newport Imports dealership, which clearly showed its franchises. On another were Lee West's name, title and contact details. On another was a map showing the dealer's location, while on another were the dealer's opening hours. It was the most impressive business card I had seen, and it had all the key details of Lee West's business.

When the dealership was built facilities were included at one level below ground level. These included a large, well-equipped gymnasium. Lee explained that this had been created not just for himself but primarily for the benefit of his customers. Customers typically only visit a car dealership once every three years to buy a new or used car and once a year to have their vehicle serviced. By offering his customers the use of a gymnasium, they would visit his dealership much more frequently, and as such become more like friends of the dealer, rather than customers. However in return for receiving this very attractive benefit free of charge the customer would be expected to buy a new car from his dealership every two years.

At ground level the premises had a facility not normally associated with a car dealership, and in this case a dealer with prestige vehicle brands. The facility was called Ruby's Diner and I suppose you could describe it as a fast-food restaurant, serving burgers, sandwiches, seafood, salads, desserts and various drinks. It was open for breakfast, lunch and dinner and had décor that included various automotive themes and artefacts. The restaurant was adjacent to the car showroom and so provided another reason and opportunity for customers and potential customers to visit Newport Imports.

As I said before a typical car customer does not visit a car dealership very often and so a car dealer's premises, which represent a very high cost of investment in buildings, are very under-visited compared to other retail businesses, such as supermarkets. Over the years I have used this dealership as an example of a car business that included non-automotive activities in its business to attract and retain new and existing customers to its premises.

After a very informative and enlightening meeting at Newport Imports, Ken McGraw drove us south to the Temecula Creek Inn where we would meet up with Range Rover customers who had paid to take part in a weekend off-road event arranged by Jack Brewer, the owner of Pioneer Centers, the Range Rover dealer in San Diego. Not only had these customers paid but they would be driving their own Range Rovers off-road, an activity that few of them had experienced before. When we arrived at the hotel, I met up with Captain Mark Phillips.

Most of the Range Rover customers were wealthy business people and I recall a few were a little concerned when we arrived in the area where the off-road adventure would take place. Mark and I shared a Range Rover, which I drove. However, before long we were in an area of rocky terrain strewn with large boulders which I would describe as extreme off-road driving. Here Mark and I both had to get out of our car and assist some of the customers. The Range Rovers could only move very slowly but by giving verbal instructions we helped them navigate their way down the uneven steep slopes, and through the rocks and boulders, without causing damage to their cars.

There was one point where I had to get in the Range Rover of one customer, a surgeon, and give him some advice and some reassurance on gear selection and in driving up a very steep incline. This gentleman had never driven off-road before and at the end of the day he said it was one of the most exciting things he had ever done.

This off-road weekend event for Range Rover customers in California proved to be very popular. Customers would experience true off-road driving, on off-road territory where you couldn't go individually. It had to be organized officially by the dealer and customers definitely benefitted from learning how to drive off-road from experts. Up until then the Range Rovers and their owners rarely left the main roads and highways of the Los Angeles area. In the case of this particular event Pioneer Centers told the customers that it would repair any car damaged during the weekend free of charge, an offer which removed some but not all the anxiety of the customers. For me it was a very enjoyable weekend, meeting many Range Rover owners and contributing in a small way to the experience many had for the first time in driving their Range Rover in real off-road conditions. On the Sunday evening I flew back from Los Angeles to the UK.

At the end of 1990 we held an end-of-year dealer council meeting at the Lords of the Manor Hotel in Upper Slaughter in the Cotswolds, a prestige venue now befitting the presence of an increasingly prestige automotive brand. It had been an enjoyable time for me in my first year as UK Sales Director and as a UK sales management team we had continued to make more progress in developing the Land Rover business in the UK for the benefit of the company, its dealers and everyone in the extended enterprise.

# Chapter 14
# Land Rover Ltd 1991

**THE YEAR 1991 STARTED WITH** a continuing and normal round of internal management meetings, which included those of a repetitive monthly nature. All the way through my career to date sales performance was measured on a monthly basis and 1991 was no different. So I held regular monthly sales meetings with the team to discuss vehicle allocations and sales performances. In addition, I held individual sales review meetings with each of the Regional Sales Managers to ensure we were achieving and, if possible, exceeding our objectives.

There were also sales demand forecast meetings, franchise operations committee meetings, franchise standards group meetings, launch timing meetings, marketing and budget review meetings. Within the Rover Group there were top management meetings on the '1991 corporate plan' called 'executive scrums' and 'distribution efficiency'. However I still managed to include dealer visits in my weekly schedule as I wanted to make sure I kept my finger on the pulse of business on the ground. It was important for me to get dealers' views on how we were doing to support their sales efforts, and to pick up their honest comments on how to improve all areas of our business.

On 1st February I attended the official opening of another new exclusive Land Rover dealership, this time Riders of Falmouth, Cornwall. This opening was supported by Chris Woodwark and the guest star was the TV and radio personality Noel Edmonds. That evening I also met Martin Peters, another member of England's 1966 World Cup-winning team.

Later in February I attended another new exclusive dealer opening with Captain Mark Phillips. This time it was Chipperfield Garage, near Watford, Hertfordshire. Once again Mark and I spoke to the assembled gathering of customers about the Range Rover team and Land Rover business respectively. Afterwards I was approached by a customer who started complaining about the Range Rover CSK he had bought recently from this dealership. He was quite upset and I expected to hear about some quality problems on his car. However, that was not the case. He told me he had been promised the lowest registration number out of the 1–200 numbers available but was angry when he found out another customer of the dealer had bought a CSK with a lower number than his. I sympathized but said there was nothing I could do.

In April John Russell informed me of a major reorganization and restructuring that Chris Woodwark and he were planning for the Land Rover Commercial organization. In practice we would be merging the existing Sales, Service and Parts field teams into one field business team. New Regional Business Managers would have full accountability and responsibility to make all decisions in their region, apart from the hiring and firing of dealers. To make this work in practice the persons chosen for this role would have to be of senior manager calibre, and we would grade the job accordingly.

This reorganization brought additional responsibilities for me in the UK and with a new title, Director UK Operations. My new role required me to carry out and manage a significant restructuring of various UK commercial departments, including my UK Sales department. I had to consider the optimum team structure and also consider and decide the best people to be in the team. As a result the changes would impact on many people in my UK Sales team and in other operational teams. During April I had a number of meetings with John Russell and John Cooper, in order to start the reorganization process.

John Russell's main message to me was that I would have to delegate even more to my first line managers. I was aware that I liked to know most of the main things that were going on. Now I accepted that in my more senior role this was not necessary. However the message I passed on to my managers was that I did not want any business surprises and that I wanted to know of any actual or potential 'bad news'.

I think it is useful to reflect on the significant changes in the organization structure that were taken by top management to drive the Land Rover business forward and the fact that unusually there was a detailed explanation as to why the changes were being made. Since joining Land Rover these changes were the most far-reaching I had experienced. So for posterity and for anyone interested in the organizational evolution in Land Rover what follows are the words contained in the Solihull Management Brief and the Rover Group communication of May 1991. I believe the communication is self-explanatory in view of its comprehensive content.

> *Land Rover Commercial Organization. Within the context of the Rover Group objective to provide extraordinary customer satisfaction, Land Rover Commercial has undertaken a thorough review of its processes. The review has established the optimum means of interface with the new Rover Group Product Supply organization and has identified the best means of progressing towards the Land Rover Commercial mission – 'We will provide Land Rover customers with outstanding products and an excellent ownership experience through the best distribution service in the world'.*
>
> *Land Rover Sales and Marketing. The review has identified a new approach to the Sales and Marketing organization based on three fundamental principles.*
>
> *NETWORK MANAGEMENT. The creation of a self-reliant dealer organization applying the process improvement philosophy embodied in TQI (total quality initiative) is a prerequisite to achieving our aim of world-class distributor and customer satisfaction. The introduction in the UK later this year is the means*

by which we will focus dealer management on the essential people and process issues to improve performance. A new position, Regional Business Manager, will be responsible for all aspects of the company interface with the network, covering Vehicle Sales, Service and Parts. The Regional Business Managers will report to the UK Operations Director and will be the prime contact with the network, able to utilize specialist Service Technical and Marketing support as required. Sales accountability and responsibility for Parts continues to rest with Land Rover Parts organization and the Regional Business Managers will provide a field service to Land Rover Parts on all Dealer Parts-related matters.

BRAND MANAGEMENT. The role of Marketing Plans for Land Rover Vehicles will be expanded to introduce the concept of Brand Management to the organization. Organized on a product line basis, the brand manager will be responsible for product line performance worldwide and will interface with the appropriate Product Director in the Product Supply organization.

CUSTOMER SERVICE. A new approach is introduced to provide Land Rover vehicle customers with a single point of contact for all enquiries. The customer service activity will enter into contracts with all areas of the company that provide support, information and services to ensure world-class standards of customer service and responsiveness. The specialist Service Technical support role will be incorporated into this new function.

The following new appointments are announced reporting to John Russell, Sales and Marketing Director; each will be the subject of further detailed announcements in due course.

UK Operations Director – John Sparrow
Export Operations Director – Chris Langton
Special Vehicles Director – Alan Edis
Customer Service Director – Peter Wyhinny
Brand Manager, Range Rover – Helen Hughes

> Brand Manager, Discovery – Colin Green
> Brand Manager, Defender – Mike Gould
> Other reports to John remain the same.
>
> Land Rover Commercial Operations. Reporting to Brian Purves, Commercial Operations Director, will be the following new appointment: Overseas Development Director – Alan Guest.
>
> Alan will use his experience, both within Overseas Manufacturing and Export Sales, to develop worldwide strategy and negotiate new KD (Knock Down) business relationships, capitalizing on the significant business opportunities which have been identified.
>
> Jim Newall will become responsible for providing the full range of Personnel services to Land Rover Commercial in addition to his responsibilities for Rover Cars Commercial.
>
> C. J. S. Woodwark, Managing Director, Land Rover.

On 29th May 1991 John Russell sent a letter to all UK Land Rover dealers outlining the new Sales and Marketing organization. John confirmed my appointment as UK Operations Director. John added the UK objective would be 'to develop a self-reliant dealer organization capable of producing the highest levels of business performance and customer satisfaction'.

> To provide the most effective and business-focused means of supporting dealers in realizing this objective we have created a new position, Regional Business Manager, reporting to the UK Operations Director. The position broadly integrates the field responsibilities of the existing Regional Sales, Service and Parts Managers, and will be the prime contact with the dealer network. The Regional Business Manager will be a senior Land Rover Manager in our organization and will be fully accountable and responsible for all field activities in his region. In particular the overall development of the business opportunities in each dealership will be a main objective.

John Russell's letter also confirmed the following appointments not mentioned in Chris Woodwark's Rover Group communication of 17th May.

→ Marketing Director – Roger Charlton
→ Commercial Director, SVO – Roland Maturi
→ Government and Military Operations Director – George Adams

Following the organization announcement of 17th May I spent some time with John Russell and John Cooper considering an organization structure for UK Operations. I was made aware of the need to work to a reduced total number of persons in achieving our optimum position. In my experience each of the previous reorganizations within the Rover Group required some 'rationalization' and my reorganization would be no different.

In order to demonstrate that these personnel decisions were being treated seriously I spent time with John Cooper interviewing existing members of the UK Sales, Service and Parts teams to assess their potential suitability for roles in the new structure. This process continued into June and July with Jim Newall and Fiona Tordoff of Commercial Personnel and I had interviews and discussions with many other people from within Land Rover Commercial and some potential candidates in Range Rover of North America.

A break in my work on the reorganization came in June 1991, when Chris Woodwark, John Russell and I hosted a group of UK dealer principals whose dealerships had been winners of the 1990 'Customer Satisfaction' incentive campaign. Together with their partners we flew from Heathrow to Paris and that evening we had dinner together on the Bateaux Mouches, which are excursion boats giving visitors a view of Paris while travelling along the River Seine.

At the Meridien Hotel at Montparnasse we changed into black tie for the next stage of our overall trip. At 8 p.m. with the men and ladies suitably dressed we travelled to the Gare de l'Est train station where we boarded the world-famous Orient Express for our overnight journey to Venice.

This should have been one of my most enjoyable trips, but during the dinner on the train I started to feel unwell. I was looking forward to having a magnificent dinner with fine wines, but it was not a pleasant experience for me. Also worrying was the fact I was a host for the Land Rover dealers when for the next twenty-four hours we should all be enjoying ourselves. I can tell you being unwell on a train is more unpleasant than it would be at home or in the office.

The accommodation on the Orient Express did not live up to my expectations. In our cabin we had a lower and an upper bunk bed, which during the day were converted back to what was described as 'banquette' seating. The cabin also had a small wash basin area hidden behind wooden panels. However, there was no toilet or shower in the cabin, facilities which in our case were further down the carriage. For one night I suppose it's an acceptable arrangement, but for me on the upper bunk bed and not feeling well I struggled to get any decent sleep and with the persistent rattling noise that trains make I could have been on any train.

The following day I was pleased I didn't feel any worse. While I couldn't enjoy the food or drink on offer I did enjoy the spectacular scenery as we passed through the Swiss Alps. In Venice we stayed at the Europa & Regina Hotel, overlooking the Grand Canal and a five-minute walk from St Mark's Square.

We were given a guided tour of Venice, including the Doge's Palace, and in the evening we all travelled in gondolas a short distance on the Grand Canal before having a final dinner at the Restaurant Malamocco. By then I had recovered and was well enough to enjoy it. On our last day we took a boat trip to Murano, a small island a mile north of Venice, famous for its glassmaking. We flew back to Heathrow at the end of a memorable trip – greatly enjoyed by all but a little less so by me.

Back at work my main focus was on the UK Operations reorganization. Also during July 1991 Charlie Hughes brought his Range Rover President's Cabinet to the UK. The purpose was to enable these key US dealers to visit the Solihull Factory and to see the vehicles being made. The dealers also visited Lex, Stourbridge, an exclusive Land Rover dealer, and went to the theatre in Birmingham to see the stage version of the film *Some Like it Hot* starring Tommy Steele. Following the show we all had dinner at the Bilash Indian Restaurant in Knowle, outside Solihull. Our American friends enjoyed this typically British way to finish the evening.

By the end of July the UK Operations structure, with the many personnel changes, was finalized. This was the last major Land Rover reorganization I was involved in and I would like to record in particular the names of the managers in the team who I announced as my first line reports. Extracts from my letter of 31st July 1991 to all Land Rover main dealers and dealer group heads which confirmed details of the new UK Operations organization are as follows.

> *Further to John Russell's letter of 29th May I am now able to give you information on the new UK Operations organization, and I am pleased to announce the following appointments reporting to me.*
>
> *These appointments will take effect from 1st August but August will be a transitional period to enable us to manage the change associated with the new field structure, which will go live on 1st September 1991. I would like to emphasize the Regional Business Manager will be the prime contact with the dealer network and he will be fully accountable and responsible for all field activities in his region. The Regional Business Managers will take policy direction from myself, but otherwise they will make operational decisions in their region. Regional Business Managers:*
>
> *Scotland and Northern Ireland: Andy Bruce*

Andy was previously Regional Sales Manager for the Scottish region.

North-West: Steve Westwood

Steve was previously Regional Sales Manager for the Northern region.

North-East: Mike Emery

Mike joins us from Range Rover North America, where he was Zone Manager for the Midwest States of America.

West Midlands: Mike Garratt

Mike was previously Regional Sales Manager for the Midlands region.

East Midlands: Vince Floyd

Vince joins us from Land Rover Export Sales, where he was Export Sales Manager.

North London and Southern: Bob Hester

Bob was previously Regional Sales Manager for the Southern Region.

South London & South: Kevin Beadle

Kevin joins us from Range Rover North America, where he was Zone Manager for the Eastern States of America.

South-Wales and West: Adrian Morris

Adrian joins us from Land Rover Marketing, where he was Forecasting Manager. He also has considerable field sales experience.

South-West: Phil Popham

Phil joins us from Land Rover Finance, where he was Finance Manager on secondment from Land Rover.

The operation of nine Regional Business Managers means we have re-grouped main dealers into nine regions as indicated on the attachments. The Regional Business Managers will be supported within UK Operations as follows.

Manager, Operations Support: David Carpenter

David will be responsible for providing a comprehensive and

*effective field support to Land Rover dealers and the Regional Business Managers.*

*Manager, Company Sales: George Hassall*

*George will be responsible for the sales of Land Rover products to major account customers in the UK. He will retain his current responsibility for Diplomatic and Tax-Free Sales, Used Land Rover product sales and Royal Household liaison.*

*Manager, Business Development: Rob Pugh*

*Rob will be responsible for the development of operational programmes and action plans, and ensure the successful implantation of the company's distribution efficiency objectives in the UK.*

*Manager, Franchise Development: Barry Tanser*

*Barry will continue to be responsible for all UK franchise development programmes, and the implementation of Land Rover corporate identity, facility development projects and franchise décor schemes. He will also be responsible for franchise planning and the development and implementation of Land Rover UK franchise strategy.*

*I am confident the new UK Operations team will be able to provide the level of assistance and support you require in enabling us to achieve our mutual objectives. I would be grateful if you would ensure all your Land Rover staff are made aware of the contents of this letter.*

*Yours sincerely,*
*John A. Sparrow*
*Director, UK Operations*

There were 125 main dealers in the UK Land Rover network at the time of my letter, and with between twelve and seventeen dealers each, the Regional Business Managers were scheduled to spend at least a full day each month with the dealer principals and the dealers' departmental management. However, the network was told we were looking to develop self-reliant dealers and so in practice dealer management was largely left to get on with the day job. Previously dealer management often complained to Land Rover senior management of having to spend too much time on the frequent visits from the three Regional Sales, Service and Parts Managers.

During September and October of 1991 there are many references in my diary to 'TCS'. This was the term to describe 'Total Customer Satisfaction', a major new operational initiative within Rover Group. TCS included specific processes aimed at delivering a consistent yet high standard of customer service that met and exceeded customer expectations. In early September TCS was introduced to the new Regional Business Managers, and then to UK Land Rover dealer principals and all UK dealer management.

I later came to realize delivering TCS and having measures that reported that customers were totally satisfied did not necessarily mean customers would be loyal to the brand. As a customer you could be very satisfied with the vehicle and the brand but next time you just wanted a change of vehicle.

On 12th November Chris Woodwark, John Russell, Roger Charlton and I hosted a UK Land Rover Dealer Incentive trip to Jamaica for twenty dealer principals and their partners. After a long journey we arrived at our hotel, the Plantation Inn in Ocho Rios. Shortly after John Russell told us he had challenged the local cricket team to a match. The cricket generated much excitement locally and also in the hotel where one very tall young waiter said he was a fast bowler in the team. I spoke to all Land Rover dealers about their willingness to play and posted a message on the hotel notice board saying, 'Cricket – your captain country needs you. Please sign up below.'

I posted details of the Land Rover team on the hotel notice board. It was clear we were unlikely to win so to have some fun and create some doubts in the minds of our opponents after each of our team's name I put the county they came from in brackets. So it gave the impression we were players from counties such as Yorkshire, Warwickshire, Surrey etc. John Russell was team captain and Ian Farnell of Farnell Land Rover, Guiseley, Yorkshire was team manager.

On 15th November in a match of twenty overs per team, Land Rover batted first. I opened the batting, but was quickly bowled by a very fast ball I didn't see! We lost a few more quick wickets before the Ocho Rios team took some pity on us and introduced their slow bowlers. I should say at this point that we had to borrow bats and two protective 'boxes' from our Ocho Rios friends. This meant each time one our team was 'out' that there was what we called a ceremonial handing over of the 'box' to the incoming batsman. However, I well remember Dennis Bunning from Stafford Land Rover, who was one of the last in our team to bat, saying he would risk batting without the box as it had been used to protect the private parts of most of the team before him.

We were all out for 105 runs, which was a fair score.

The opposition's captain and opening batsman looked like Viv Richards, the famous West Indian cricketer. He was soon scoring hard-hitting boundaries. However, he didn't last long. Brian Horsnell from Webbers of Basingstoke bowled a short ball and 'Viv' hooked it hard at head-height towards me, fielding at square leg. I instinctively put my hands up to my head and much to my amazement the ball stuck firmly in my hands and our Viv Richards lookalike was out. There was much jubilation in our team for a few brief moments at this small victory. It was not long before our opponents inevitably won the match when they reached the total of 106 for the loss of four wickets. After the match John Russell the Land Rover captain presented the victors with a trophy already engraved with their name!

Thus 1991 was a significant and memorable year for me in my career in Land Rover.

Chapter 15

# Land Rover Ltd 1992–1993

**ON 1ST JANUARY 1992** I received a letter from the Ladbrook Park Golf Club inviting me to attend an interview at the club at which I was supported by Geoff Dale, Managing Director of the Evans Halshaw Motor Group and a member of the club.

I explained to three gentlemen officials of the club at the interview my wish to join a high-quality local golf club, my interest in golf and my fair golfing ability, referring to the latter by my play in a number of pro-am tournaments. I added that as a Director of Land Rover I had a responsible position in the Solihull area. I also lived in Dorridge, not far away.

Towards the end of the interview I was asked if I had any questions. In reply I simply said I would like to play the golf course before deciding. There was a look of surprise on the faces of the three officials to what I said. Geoff gave me a kick in the leg under the table and said I shouldn't worry about that and he would arrange a game at the club shortly. It seemed a reasonable question for me to ask. To me it was the equivalent of buying a new car. You would want to have a test drive before deciding whether or not to buy.

I had to complete a document saying, 'I desire to become a member of Ladbrook Park Golf Club Limited.' I was proposed by Geoff Dale, and seconded by Anthony Archer, also a Director of Evans Halshaw and a member of the club. However it was not until May 1992 that I had my first round of golf at the club, with Geoff. I have continued my membership since moving to Oxfordshire in 2001, because I greatly enjoy playing this excellent golf course.

During March and April 1992, I had a number of meetings with John Russell regarding my next job. John told me I had achieved a great deal for Land Rover in the UK and he now wanted me to use my knowledge and experience in major international markets. I was very happy in my Director, UK Operations role, but I was to become Director, Business Development, a new position responsible for the development of Land Rover's business worldwide.

In March 1992 John had been to Australia with other Directors to review the Land Rover business situation as the existing management buyout from the previous BL Australia organization was in trouble. What resulted was Project Wombat, a team to re-establish a Rover Group wholly owned National Sales Company, concentrating on the Land Rover business. I was to become very involved in this project.

At the same time the development of Land Rover Centers in the USA had reached an important stage. Charlie Hughes, President of Land Rover of North America, was planning to hold a series of dealer meetings across the country to present the latest plans for Land Rover Centers and I would attend these to explain how we had successfully developed a network of exclusive and totally dedicated Land Rover dealers in the UK. In my presentation I would give details of dealer investment levels, franchise standards, levels of customer satisfaction and franchise profitability. All this would be aimed at demonstrating to the US dealers the business benefits of following a similar direction to the Land Rover UK dealers.

Charlie had confirmed dealer meetings at seven locations across the USA in the four-day period 20th–23rd April 1992. To achieve this meeting schedule would have been impossible using normal airlines. However, Charlie had arranged for us to travel using the British Aerospace (BAe) demonstrator, a BAe 125-800 small twin-engine version of the corporate jet. I think the jet had a capacity of eight passengers and two pilots. We would also have two pilots but there would be just five of us on this trip: Charlie Hughes; Roger Ball, Vice

President Operations; Dave Schworm, General Manager Service; Monica Quagliotti, Manager Special Events; and me.

On Sunday, 19th April Tony Cottrill the Land Rover chauffeur picked me up at home and drove me to Heathrow where I took the BA flight to Washington Dulles Airport. The first of the seven dealer meetings the following day was held at the Hyatt Dulles Hotel at Washington Airport. Starting at noon the meeting ran until 2 p.m. At the internal area of the airport, we took our first flight to San Francisco. It was a unique experience to travel in a way usually just for the rich and famous and I remember it was a very enjoyable journey on this executive jet.

At San Francisco Airport we were picked up by the Regional Manager and being able to avoid all normal flight arrivals procedures, were taken direct to the Hyatt Regency Hotel at the Airport. This would be the venue for the second dealer meeting the following morning starting at 9 a.m. and finishing promptly at 11. After the meeting had finished, we were soon back on our jet to Los Angeles Airport where at 2 p.m. we had our third dealer meeting at the Marriott Hotel. The two-hour meeting finished at 4 p.m.

We now took our next jet flight to Dallas, Texas and stayed overnight at the Stouffer Hotel. This was the location for our fourth dealer meeting, starting again at 9 a.m. and finishing at 11 a.m. Once this was over, we flew to Chicago O'Hare Airport where we held our next dealer meeting at the Hyatt Regency Hotel. This started at 3 p.m. and two hours later meeting number five had been completed.

By now the presentations had taken place so quickly one after the other that all of us were pleased they were running well and being well received. For me it was as if we were on our equivalent of a theatrical tour, and at each venue my colleagues and I were on stage in front of our dealer audience.

After the Chicago meeting we flew to Atlanta, Georgia where we stayed overnight at the Hyatt Hotel at Atlanta Airport. It had been a very busy day starting in Dallas, and via Chicago to Atlanta. Day four was to be our final day of dealer meetings. Meeting number six was held at the Hyatt Hotel in Atlanta and the two-hour session finished at 11 a.m. we flew to White Plains, near New York where meeting number seven would be held at the Crowne Plaza Hotel starting at 3 p.m.

It was at this point that I was aware my very busy week was not going to end with my flight back to the UK. I also knew that on 23rd April back in the UK, an announcement was made to UK Land Rover staff regarding organization changes, which included my new position as Director, Business Development.

At 4.15 p.m., having made my presentation I made my apologies for leaving the meeting early explaining I had to catch a specific BA flight at 7 p.m. from John F. Kennedy Airport to London in order to be in time to host a four-day UK Land Rover dealer management incentive trip to Euro Disney, which involved an early-afternoon flight from Birmingham to Paris. My early departure from the dealer meeting was indeed necessary as I remember there was considerable traffic that afternoon on the 35-mile journey from White Plains to JFK Airport which took over an hour, and I was very anxious at times when the traffic didn't move.

My BA overnight flight to London Heathrow arrived at 7 a.m. on the Friday, and I was pleased to see Tony Cottrill waiting for me. He got me home a couple of hours later. I had just enough time to unpack my suitcase and repack it with some fresh clothing for the four-day trip to France. By late morning we were meeting the UK dealers at the Land Rover Driving Experience facility in Solihull and at 2 p.m. we boarded our Air France flight to Paris at Birmingham Airport.

Euro Disney, now called Disneyland Paris, is located about 20 miles east of Paris. It had only opened for business and to the public on 12th April 1992, just twelve days before our Land Rover party arrived. In fact we were told we were the first corporate group to visit Euro Disney. We would be staying at the five-star Hotel New York, located within the Euro Disney Park. On our first evening we were entertained by Buffalo Bill's Wild West Show. I don't remember much about that as I am sure I was probably still suffering from some jetlag and tiredness.

The next day was Saturday and we spent our time mainly within Euro Disney experiencing some of the attractions. I remember taking a photograph of my wife Annette outside 'Annette's Diner' and that it was not very warm as we toured the park.

By midnight after a gala dinner in Paris we were at the Moulin Rouge (Red Mill in English), probably best known as the birthplace of the modern form of the can-can dance. We were seated literally in the front row adjacent to but below the stage where the scantily clad young ladies danced and performed. It was interesting to say the least to be looking up at the dancers, who at times were above us in touching distance!

On 27th April we travelled back to Birmingham Airport and that night I was pleased to be sleeping in my own bed for a change. That week I noted in my diary that there were nine regional dealer meetings in the UK to present details of the Discovery Customer Commitment Programme. So early on Tuesday morning I was on the plane from Birmingham to Glasgow to present the programme to the Scottish dealers. That night I was back home and the following day I travelled a short distance by car and my presentation this time was to the West Midlands area dealers in Solihull.

That month I had a number of meetings with John Russell to discuss my Business Development job and I was involved in various UK business reviews and dealer group review meetings as part of an orderly and gradual handover to Peter Wyhinny.

At the end of May I made the first of many visits to the Rover France national sales company, located at Argenteuil, on the northern edge of Paris. It was a two-day business review meeting with Chris Franklin, the Managing Director, and his management team, including Steve Norman, Marketing Director, and Claude Dehon. On the afternoon of day one we had a franchise review, while on day two we covered Rover France's business plans, including product, sales performance, vehicle marketing and dealer network. British Airways had a good schedule of flights from Birmingham to Paris Charles De Gaulle Airport so travelling was fairly convenient at that time. Unfortunately British Airways no longer flies in or out of Birmingham Airport.

Having not long returned from my business trip to the USA I was aware that fairly soon the activity to develop Land Rover Centers in the USA would speed up. So on 2nd June 1992 I attended the first meeting of Project Team Columbus at our offices in Bickenhill, near Birmingham Airport. Charlie Hughes had chosen the word Columbus to remind the team to rediscover America. Though Christopher Columbus was not the first European explorer to reach the Americas (having been preceded by the Viking expedition led by Leif Ericson) his voyages led to the first lasting European contact with the Americas, inaugurating a period of European exploration, conquest and colonization that lasted for several centuries.

Team Columbus would be a multidisciplinary team and would comprise the following personnel.

Rover Group UK: Terry Morgan, Land Rover Manufacturing Director; Graham Broome, Manufacturing Director Rover Power Train; Harry Reilly, Finance Director Land Rover; Garry Smith, HR Manager Land Rover; Alan Stedall, Land Rover Parts Systems Director; Keith Taylor, Land Rover Parts Director; and me, UK Operations Director Land Rover. I may have got some of the UK team's job titles wrong, but it was over thirty years ago.

Range Rover North America (RRNA): Charlie Hughes, President; Joel Scharfer, Vice President Finance & Administration; Roger Ball, Vice President Operations; Sally Eastwood, HR Manager; Rainer Freuchnicht, Vice President Sterling Motor Cars (SMC); and RRNA subject fact holders.

On 9th June 1992 Charlie Hughes sent a memo to the Columbus members setting out the initial objectives for Team Columbus. It marked the start of a significant company initiative for the Land Rover business in North America and so for posterity I reproduce Charlie's memo as follows.

> We are looking forward to working together and to breaking new ground regarding the best way for Rover Group to realize the potential that exists for our products in North America.
>
> To develop a three-year (1993–1995) business plan that examines the potential in North America for Land Rover products, to determine what we are committed to achieve, to agree on the resources that are required, and to gain the commitment of this plan by all the involved parties of Rover Group.
>
> To use Team Columbus as an opportunity to develop a process for examining the potential of mature markets and the development of a cogent business plan.
>
> With six working days to complete this task, we will have to be organized and time-efficient, for the development of this type of business plan will require the examination and debate of much in market analysis. It will also require agreeing on some hard targets in a way not done previously.
>
> Attached is an initial agenda. The process will include a break-out of key subjects into modules. Each module will be comprised of a presentation of factual information by the appropriate responsible parties leading to a discussion and decision among Team Columbus members. Because of the complexities of some of these matters and to allow team members to do some normal work we will schedule two modules per day, one in the morning from 9:00 a.m. until 12 noon and one in the afternoon from 2 p.m. until 5 p.m. As we complete each module we will keep a running score card of what was decided in each section, knowing that new information from succeeding sessions may lead to a need for rediscussion of a previous point and a possible change of decision. We will allow for this discussion each evening starting at 6 p.m. and going until completion or 8 p.m., whichever comes first.

*The most critical of these sections will cover price and volumes. That is scheduled for Wednesday and it must be resolved that day no matter how long it takes.*

*We are planning to be finished at close of business on Friday and during that afternoon we must allow time to determine how we will cascade this plan throughout Rover Group.*

*While all team members are committed to making this process work and putting in as much time and energy as is required, we must also stay as fresh as possible, maintain a realistic perspective and avoid as much as possible ending up with a bunker mentality. In that regard, we believe it is important that all Team Columbus meetings be held with the entire group present.*

*Discussions of evolving the process should occur during our evening sessions. We have the opportunity to 'write the book' on the subject, so we must allow ourselves to learn as we go.*

*We look forward to getting together on Saturday night, 20th June, and kicking off our first session on Sunday. This is an exciting chance to not only seize what we believe is a sizeable opportunity available to Rover Group in North America, but also help develop a process that will allow us to grow our business throughout the world.*

The Team Columbus Agenda and Modules for the week were as follows.

*Monday: Quality/Warranty/Goodwill (Lemon Law)/Cost of Ownership. Product/Model Line-Up.*

*Tuesday: LRC (LR Centers)/Dealerization. Advertising. Parts. Discovery Research.*

*Wednesday: Marketing/Corporate Communications. Price/Volumes.*

*Thursday: Organization/Facilities/Logistics.*

*Friday: Plan Finalization. Next Steps.*

At 6:30 a.m. on Saturday, 20th June I drove to Gatwick Airport where we all met up to catch the US Air 10:30 a.m. flight to Baltimore. Once we arrived there at 2 p.m. local time we were picked up and taken to the Ramada Hotel in Annapolis, where we would stay for six nights.

On Sunday, 21st we had our first team meeting at RRNA's offices in Lanham. The agenda covered Team Columbus objectives, the project process, RRNA values, mission and objectives, all presented by Charlie Hughes. Roger Ball presented Project Eagle, Joel Scharfer presented RRNA Financials, Rainer Freuchnicht presented SMC Financials, Jim Lehmann presented Government Factors.

We made a lot of progress on day one which left some time in the afternoon for us to visit a Lexus dealer and a Saturn dealer, both in Rockville, a suburb of Washington. The team worked well and following the original agenda we stayed on track and on time. I don't recall us having to work late into the evenings and on the Thursday evening our dinner comprised of 'crab cracking' at a restaurant by the sea in Annapolis. This was an interesting experience, not having done this before, but it was a messy way of producing your evening meal. By the afternoon of Friday, 26th June an Action Plan was agreed.

We were able to leave Lanham and our friends in RRNA in time to catch our scheduled night flight from Baltimore to London Gatwick. We arrived there at 8 a.m. and I was home by 11:30 a.m. In the week after we returned the UK members of Team Columbus met in Terry Morgan's office to ensure all team members' views and experiences were captured. We also discussed and prepared the content of the team report. I had already drafted my part of the report. It was not until the end of July that the summary report was produced, the content having been coordinated by Graham Broome. It was not circulated however until I returned from a very long business trip to Australia on 18th August. I have decided it is appropriate that I record in detail here the main points of the Team Columbus report, as contained in the two-page report summary.

The summary starts by outlining the UK and RRNA members of Team Columbus as detailed earlier in this chapter. The rest of the summary is as follows.

*The team reviewed all aspects of the Range Rover of North America business in a working environment which was committed, open, honest, frank and positive. The review process resulted in the unanimous recognition that North America is a market of significant potential which has to be exploited further if Land Rover is to be successful.*

*The team agreed upon three unshakeable facts:*

*1 – The US is the largest and toughest 4x4 market in the world. Large: 1 million vehicles per annum (57% of the world). Tough: fifteen makes, twenty-seven models.*
*2 – To differentiate Land Rover from its competitors we need to establish Land Rover Centers as a unique selling proposition, thus following a well proven route, i.e.: UK Land Rover experience; US Lexus, Saturn experience; US Land Rover solus dealer (1) successful experience.*
*3 – Discovery volume is critical to the successful establishment and development of Land Rover Centers, and in essence 'For Land Rover to succeed, it must be successful in the US'.*

*To achieve success in the US a grouping of seven critical success factors (CSFs) emerged, as given below.*

*1 – Competitive product quality is required.*
*2 – Extraordinary customer care will be necessary, delivered through Land Rover Centers (LRCs).*
*3 – Awareness of marque values must be established through effective targeted marketing to achieve required impact.*
*4 – Discovery must be introduced in 1993.*
*5 – Products of competitive price and specification that recognizes the premium value of the marque must be offered.*
*6 – Waste must be eliminated between Rover Group and the NSC through the improvement of interface processes.*
*7 – Extraordinary customer satisfaction must continue to be developed throughout the SMC withdrawal.*

*The discussions which took place during the week at Lanham have resulted in a number of agreed actions which will ensure the CSFs are delivered. These are detailed in section two of the report. [There were thirty-one agreed actions].*

*The team departed the US leaving behind an NSC which owned and was committed to deliver a growth plan with an increased contribution potential for Rover Group. The plan to achieve sales of 10,500 units in 1995 was considered to be totally robust by the whole team. The NSC team was also committed to deliver the plan with a leaner organization.*

*It was recognized that the review process was equally applicable to other National Sales Companies (NSCs). Sections three, four and five of this report summarize in a 'plan on a page' format – the process actually adopted; the process improvement recommendations; and a blueprint process recommendation.*

I have looked through my copy of the Team Columbus report several times in writing this section and I consider what it contains is an excellent template for any major automotive project, even today, over thirty years later. It is not appropriate to record all thirty-one agreed actions here, but I know Team Columbus and the action plans it generated did provide a firm foundation for the continued development and growth of a successful Land Rover business in North America. I am just very pleased to have been part of that project team, to have made my contribution to the input and outcomes, and to have learned a great deal from that unique experience.

My overseas travel continued a few weeks later when on 8th July I attended the European and international markets launch of the 1993 model-year Range Rover in Marrakech. It was a longer flight than I expected as the Royal Air Maroc flight landed at Tangiers and Casablanca en route. Our hotel in Marrakech for the six nights was the Pullman Mansour Eddahbi. Our first day there was the ride-and-drive part of the launch for the German and Australian delegates. Annette and I drove with John and Norma Russell in one Range Rover. This was an exceptional location to experience the new Range Rover and we drove in a convoy of some thirty vehicles south through the Atlas Mountains. We were on roads most of the time and so we did not experience true off-road conditions.

In the afternoon back at the hotel I had a meeting with John Shingleton, the recently appointed Managing Director of Rover Australia. As I mentioned earlier Project Wombat's objective was to look at re-establishing a viable Rover Group NSC in Australia, based primarily on Land Rover business. I was aware that my role would be to support John Shingleton, especially on dealer network development, and that in due course I would visit Australia.

In my meeting with John we discussed a course of action and agreed that my knowledge and experience of establishing a dedicated Land Rover network in the UK would be used to motivate the Australian dealers. My recent Team Columbus involvement in the USA would also be helpful to demonstrate that Land Rover's dealer development initiatives were serious and not confined to Australia.

John said that I should visit for two months to allow sufficient time for me to visit all the dealers and to work with his team, not only the Regional Managers but the office-based management. I responded by saying I would go for a few weeks. John told me this wouldn't be sufficient to do everything that needed doing and so I agreed to a visit of up to four weeks.

We met again at my office in Solihull on 15th July. John Shingleton said my visit should take place as soon as possible. Following a discussion with my boss John Russell, it was agreed my visit to Australia was a priority and so I agreed to travel to Sydney on Saturday, 25th July. The flight time to Australia, including one stop, is around twenty-four hours so leaving the UK on the Saturday I would arrive in Sydney all being well on the Sunday evening. This would ensure that we could start work on Monday, 27th.

John Shingleton suggested I fly on Qantas QF2, to which I agreed. This flight would leave London Heathrow at 1 p.m. on the Saturday and arrive in Sydney at 8 p.m. on the Sunday, with a short refuelling stop in Bangkok. There would be no time for me to get over any jetlag but at least I would be travelling in some comfort in business class. Out of interest, do you know where the Qantas name comes from? Qantas is an acronym for its original name, 'Queensland and Northern Territory Aerial Services'.

I confirmed my flight timings to John Shingleton and received a fax from him confirming I would be picked up on arriving in Sydney. However, I was surprised that he also said my accommodation had been booked in a nice pub in Parramatta, a suburb of Sydney where the Rover Australia offices were situated. I wasn't expecting to stay in a pub!

The flights from Heathrow via Bangkok to Sydney were on time but it was a long time to be sitting down. I was pleased to have time to stretch my legs during the short stop at Bangkok airport. When I arrived in Sydney and asked about the pub I found out that I was staying at the Huntley Hotel in Parramatta. I was told that an Australian pub or hotel is an establishment licensed to serve alcoholic drinks for consumption on the premises. They also provide accommodation and other services, as entertainment venues, and restaurants serving meals. John Shingleton could easily have said I was booked into the Huntley Hotel but he was having some fun at my expense as I did not know in Australia pub could mean hotel.

On the Monday morning I received some briefing presentations from John and three of his Regional Managers, Tony Eagleton, John Morgan and Owen Peake. I was also introduced to other members of the team. The following day was arranged as a Rover Australia Franchise Committee Review which included a session to discuss and consider the market potential for Rover and Land Rover products in Australia.

It was made clear that one of the main reasons for my visit was to visit almost every dealer in the Rover Australia network, to reaffirm the Rover

Group HQ strategy of re-establishing a viable Rover Australia business and of the importance of developing a strong profitable dealer network to achieve this. During my visits to the dealers I would be discussing the need for them to commit to the Rover Australia franchises and to the relevant franchise standards. This would require many dealers, especially those in the cities of Sydney, Melbourne, Adelaide, Brisbane and Perth, to make investments in exclusive premises development.

I remember an outline map of Australia being overlaid and totally covering a map of Europe with the comment that Australia is an extremely large geographical continent and that most of my travel within the country would be by air. A detailed day-by-day schedule for my visit had been carefully developed and arranged and this was presented to me. Almost all the flights would be with Ansett, the major internal airline at the time. Even before I arrived in Australia, I was aware there would be a considerable amount of travel and when I looked through the schedule I realized I would be visiting every city and almost every town, with the exception of the capital Canberra and Darwin in the extreme north of the country.

My day-by-day itinerary, according to my diary entries, together with some details of my journey through Australia, was as follows.

29th July 1992: Dealer visits in Sydney with Tony Eagleton, Regional Manager for New South Wales. In the morning we visited Arthur Garthon Motors in Hurstville, Sydney and in the afternoon Alan and Rod Dale at Purnell Motors, Arncliffe, Sydney. This was to be the last of four nights at the Huntley Hotel. I had travelled to Australia with two large hanging suit carriers but on these internal flights I would only be able to take one. This would mean using the hotel laundries when necessary.

30th July: In the morning we had another meeting at the Rover Australia offices. In the afternoon Tony and I made a dealer visit to Ian Pagent at Asquith & Johnstone in Parramatta. At 5 p.m. I flew from Sydney to Coolangatta on the Gold Coast. Here I was picked up by John Morgan, the Regional Manager for Queensland. Overnight was at the Pan Pacific Hotel at Broadbeach on the Gold Coast. We had dinner at Mario's Italian Restaurant at Broadbeach and then visited Jupiter's Casino on Broadbeach Island. At the latter we mainly watched the large number of people playing, and mainly losing money, at the card tables. I was told there were many flights from Japan with Jupiter's Casino the final destination for the very many Japanese visitors, even though it was a nine-hour flight.

31st July: The morning dealer visit was to Jim Hill and Brian Smith at Southport Motors, Southport, Gold Coast. Lunch and afternoon with Peter Mikeleit and Alan Piper of Austral Motors, Brisbane. We had dinner at Gambaro's in Brisbane, a restaurant noted for its giant prawns. Overnight was at the Chancellor in the Park Hotel in Brisbane.

1st August: We now arrived at my first weekend in Australia. This Saturday John Morgan told me we would be travelling to Lismore to attend the Summerland Classic Car Rally held there. John together with his wife Sue and I would drive south from Brisbane to Lismore in his elderly Jaguar, a journey of around 125 miles. Taking the coastal road we stopped at Byron Bay, a beachside town. Here Cape Byron, a headland adjacent to the town, is the easternmost point of mainland Australia.

The history of Europeans in Byron Bay began in 1770, when Lieutenant James Cook, better known later as Captain Cook, found a safe anchorage and named Cape Byron after a fellow sailor and explorer John Byron, the grandfather of the poet Lord Byron. We had dinner in Lismore and stayed overnight there at the Centre Point Motel.

2nd August: We attended the Classic Car Rally where John Morgan entered his Jaguar. I don't think it won any prizes. After an enjoyable Sunday we travelled back to Brisbane and had a steak dinner at the Breakfast Creek Hotel, arguably one of the most famous and popular Steakhouses in Queensland. That night I was back in the Chancellor in the Park Hotel. While I was away that weekend in Lismore I had my first batch of laundry done by the hotel.

3rd August: Monday morning started with a 100-minute flight from Brisbane 500 miles north to the coastal city of Mackay. Mackay with a population of around 75,000 is nicknamed the sugar capital of Australia because its region produces more than a third of Australia's cane sugar. Shortly after our flight landed, we visited Dennis Roberts and Brian Park of Roberts Motors. This was a business meeting that took place over lunch because not long after we were back at the airport to take an hour's flight some 200 miles north to Townsville. This city has a population of around 180,000 and so is the largest urban area north of Brisbane. Dinner was with Rex Keen and Ron Considine at Dynasty, a Chinese restaurant in Townsville. That night I stayed at the Sheraton Breakwater Hotel in the city.

4th August: There would be two airline trips this day too. The first dealer visit was at 8:30 a.m. to Tony Ireland of Tony Ireland Holden, in Townsville. At 11:30 a.m. I was on an hour's flight from Townsville north again 220 miles to Cairns, a major city on the east coast of North Queensland. Cairns with a population of almost 165,000 is a popular travel destination for tourists because of its tropical climate and access to the Great Barrier Reef. Here I met with John and Adam Broadley of John Broadley Motors. At 4:30 p.m. John Morgan and I took our second flight of the day from Cairns back to Brisbane. That night I was back at the Chancellor in the Park Hotel, for a third night there. I was getting to know it quite well.

5th August: There would be another flight on this day but not before two dealers' visits reached by car. The first in the morning was a trip some 25 miles south-west of Brisbane to the city of Ipswich to meet Paul Newton of Blue Ribbon Motors. Following that meeting we travelled 55 miles westwards to Toowoomba, which with around 160,000 persons claims to be the most populous inland city in the country after the national capital, Canberra. There we met Allan Flohr. Later in the afternoon John Morgan drove me back to Brisbane Airport where at 6:20 p.m. I caught flight AN69 to Melbourne. Around two and a half hours later Owen Peake, the Regional Manager for Victoria, South Australia and Tasmania, picked me up from the airport and we went straight to dinner with two dealers, John Ayre and Lance Dixon, whose premises I would visit the following day. That night and for the following night I stayed in the Park Royal Hotel in St Kilda, Melbourne.

6th August: There were dealer visits in the Melbourne area. First Owen and I visited Lance Dixon at his dealership in the Doncaster district of the city. The second visit was to Stewart Webster at Frankston. The third meeting was at ULR with John Ayre, his wife Louise and Gary Brill. At the time ULR was the biggest exclusive and most successful Land Rover dealer in Australia, and one that I regarded as a role model for dealers in the other major cities. John and Louise Ayre have become good long-lasting friends as we meet up every year for a dinner when they come to the UK.

7th August: We ventured outside Melbourne to see Michael Peck and Bruce Stokes at Peck & Stokes of Geelong. This was followed by a visit to Gardon Motors at Ballarat. At 3:10 p.m. Owen and I flew from Melbourne to Hobart, the capital of the island of Tasmania. In Hobart we visited Terry Hickey's dealership and had dinner with him at Prosser's restaurant. That night we stayed at the Sheraton Hotel in Hobart.

8th August: This Saturday, and the start of my second weekend in Australia, turned out to be very memorable. Owen had arranged for some sightseeing. Leaving Hobart we drove first of all to Richmond, a small town 15 miles away. Here was the oldest bridge in Australia still in use. The Richmond Bridge was built in 1823 to 1825, around the time of the town's first settlement. I found it a little hard to believe that the bridge I was standing on was only 170 years old, but still the oldest in the country. From Richmond we continued our drive to Port Arthur, a small town and former convict settlement about 37 miles south-east of Hobart.

From 1833 until 1853, it was the destination for the hardest of convicted British criminals. If you are interested in knowing more about Port Arthur, please visit its website. Having been there and learned about the harsh conditions that convicts endured here it is not surprising to me that Port Arthur is officially Tasmania's top tourist attraction. At the time it was Australia's equivalent of the USA's Alcatraz. Not the place to be sent!

Another flight took me from Melbourne to Adelaide, the state capital of South Australia. There I met Bob Corradine and Rudi Aharmer of Prestige Solitaire, and visited their dealer premises in the city districts of Hawthorn and Walkerville. Unfortunately the schedule did not allow more than half a day there, and at 6:30 p.m. I was on a flight to Perth, the state capital of Western Australia.

I had been told to dress casually for the meeting with Alf Barbagallo, the owner of the Perth dealership of his name. John Shingleton, the MD of Rover Australia, and Tony Eagleton had also travelled to Perth for the meeting, because the three of us were aiming to get Alf to commit to invest in and develop an exclusive Land Rover/Rover dealership in Perth. This if successful would have a significant impact on Rover Group sales because to succeed in Western Australia you had to succeed in Perth.

At 10 a.m. we arrived at the dealer premises for our meeting with Alf. However, much to my surprise, Alf said we would be doing something else first. We travelled out of Perth to the Wanneroo Motor Racing Circuit, 30 miles north of the city. This was certainly a big surprise and not what I was expecting. I didn't know Alf Barbagallo had been an Australian Touring Car Championship driver. Alf won six Western Australian Sprintcar Championships between 1967 and 1982.

At Wanneroo Alf said he planned for me to experience a few laps in his Holden Commodore Number 77 saloon race car. That was another surprise. Alf had brought his car racing team there and joining us that day was Ian Love and his racing saloon car and his race team. That was another surprise. It had been raining and the race track was wet. So Alf decided the two cars would do a few laps to determine which tyres would best suit the track conditions.

It was approaching lunch time and a barbecue had been organized for us. Following lunch Alf said it was time for a race. I would go with Alf and John Shingleton would go in Ian Love's car. Alf and Ian would be driving the cars they drove in an Australian saloon car racing series. Alf's Holden Commodore and Ian's equivalent car had 560 bhp and a top speed over 150 miles an hour. Before we set off Alf and Ian went out for a few laps. Then I was seated and strapped into the passenger's seat on the left-hand side of the car and given a crash helmet. We were ready to go and Alf said we would have a two-car race. We would do one practice-type lap and if towards the end of it I put my thumb up we would start the race. I assumed Ian Love and John Shingleton had no option but to follow Alf when we 'went for it'!

It was an incredible experience. Alf's son Troy and Alf's brother Tony said I should watch Alf's driving and especially his feet when changing gear. So I did. The speed of his foot movements was very fast and clearly something professional motor racing drivers have to do because with a manual gearbox and rapid gear changes up and down this speed of movement is essential. I remember the incredible acceleration out of corners and the very rapid deceleration from heavy braking as we approached sharp corners. Initially I felt we were approaching corners much too fast to have any chance of getting round them safely, but Alf's skill in keeping the car controlled under severe braking was very impressive to watch.

The race itself was one where at all times, if you excuse the words, one car was right up the backside of the other. That wasn't worrying when we were in front but on the few occasions when we were right behind Ian's car I was sure a crash was possible. But it didn't happen and we finished in front when the race was over.

When we emerged from the car Troy said we had set a fastest lap of 64.7 seconds compared to the lap record at the time of 61.3 seconds over the circuit distance of 1.5 miles. He told us Alf and Ian had treated the 'race' fairly seriously. Not only was it an exciting experience but I realized I could never have been a motor racing driver.

After a day motor racing we travelled back to Perth. That evening we had an excellent dinner with Alf and Ian at Coco's, one of Perth's premier restaurants, owned by Ian and situated on the banks of the Swan River on the South Perth Esplanade. It was great way to end the day.

12th August: at 9:30 a.m. we arrived at Alf's car dealership for the important business meeting I thought we were going to have the previous day. We had got to know each other quite well during the non-business hours together and during the morning meeting's discussions Alf agreed to make the necessary investment we were seeking in Perth. It was a visit I shall not forget.

John Shingleton, Tony Eagleton and I would be flying back to Sydney in the afternoon after our meeting with Alf Barbagallo. I was expecting to spend a few more nights at the Huntley Hotel in Parramatta, but I said to John it was a great shame that I had not seen any of the sights of Sydney, including the Sydney Opera House, during my time in Australia. So John arranged for my hotel booking to be changed so that I could spend the last few nights in the centre of Sydney. I was now booked into the luxury Park Hyatt Hotel which overlooked the Sydney Harbour Bridge and the Opera House.

The trip of just over 2,000 miles from Perth to Sydney was the longest internal flight I made during my time travelling in Australia. It took four and a half hours as we flew into a head wind much of the way. This particular journey and my earlier flights certainly reminded me of the large geographical size of the continent.

I was fortunate to have a room at the Park Hyatt Hotel so that when I opened the curtains in the morning the Sydney Opera House was there straight in front of me across Sydney Harbour. This was one of the most memorable sights that I have ever had from a hotel room during my business travels. I think it is one of the best-situated hotels in the world.

That afternoon I returned with Tony to the Rover Australia offices in Parramatta. Our meeting there included a report and presentation I made to John Shingleton and his senior management team of John Skinner, David Watson and Bob Dart. I was reassured that the dealers I visited had all agreed to make the investment in Land Rover we were seeking in order to grow the business on a sustainable basis. They had been informed that a new set of franchise standards would be introduced and that they would receive full support in developing their businesses.

Fortunately the next day there was no more business to be conducted so I had time to do a little sightseeing. This included a leisurely walk around Sydney Harbour, the area around the Opera House and the shopping area. In the afternoon I took a taxi from the hotel to Sydney Airport and at 4 p.m. I boarded Qantas flight QF1 to London Heathrow via Bangkok. My flight arrived back at Heathrow at 7:20 a.m. on Saturday, 15th August.

I am sure my visit to Australia had a beneficial effect on the subsequent work of the Rover Australia team in developing their franchise strategy and dealer development plans. I was able to demonstrate the success of Land Rover's dealer-related activities in the UK and that Land Rover in North America was following a similar path. In discussing with the dealers our similar plans for Rover Australia I am sure this did help in raising the confidence levels of all the twenty-four dealers I visited in their own locations.

I think most of the dealers I met were pleased and reassured that I had taken the time to make a visit to their dealership. One of the things I remember most about my dealer meetings was the comment that almost all of them made to me towards the end of my visits to their premises. It was, 'We don't suppose we will ever see you again.'

I think this comment was made on the historical evidence that few if any Rover Group Directors from the company's UK Head Office ever made a return visit to a dealer's premises in Australia. I could only reply by saying that if I was still doing the same job and had the opportunity to revisit Australia I would aim to make some return dealer visits. Less than four years later I was fortunate to be able to return to Australia for a major new Range Rover model launch which did enable me to meet almost all of them again.

Back home I had a few days in the office before being involved in the visit of the Land Rover of North America dealers to the UK. Their visit commenced on 23rd August and finished on 27th August. I should say here my wife Annette had a unique sense of humour at times. The visit of Land Rover dealers ended with a black-tie dinner at Blenheim Palace. We were introduced just before entering the dining area. A man asked the couple's names. When asked for ours Annette replied Mr and Mrs Dan Quayle, at that time the Vice President of the USA. I thought Annette's joke might have been a career-limiting comment. Fortunately not.

When markets such as the USA and Australia were starting out to develop exclusive Land Rover dealer networks visits to UK dealers that had already made significant investments in the Land Rover franchise were invaluable in encouraging and convincing others to follow.

On 18th October 1992 I flew to the USA again. This time I would be joining Charlie, David Schworm, Bill Baker and Ken McGraw and Land Rover cabinet members in a two-day event at the offices of Design Forum in Dayton, Ohio.

Design Forum, headed by Lee Carpenter, was one of the country's top retail design firms and it had been selected to help research, design and develop the interior and exterior features and architecture of LRNA's Land Rover Centers. Our few days with Design Forum would largely be spent in a very comprehensive 'carding session' on Land Rover Center development. This session involved putting all ideas on all elements of the Land Rover brand on cards, including all ideas on how these could be incorporated into a unique best practice Land Rover Center. Every idea suggested, however extreme or crazy, was captured as part of what some would call a 'brainstorming' event. At the end of this session we left Lee and his team to distil the ideas into a first proposal for the Land Rover brand and a Land Rover Center.

Days later with Stan Ford the Regional Manager driving we travelled to La Casa del Zorro in Borrego Springs to attend and support the 'Bauer Wheels Event – Anza-Borrego' organized by Dick Bauer's Anaheim dealership. Anza-Borrego Desert State Park (ABDSP) is a state park located within the Colorado Desert of southern California, and includes 500 miles of dirt roads and off-road track, and twelve designated wilderness areas. It is around a two-hour drive north-east from San Diego, and south of Palm Springs. The event would give Bauer's Range Rover customers the opportunity to drive their Range Rovers off-road and with some extreme off-road conditions. Most of these people only drove their Range Rover on the highways and roads of California, so this weekend event would be something most of them had never experienced.

The event was something the dealer could arrange because the severe off-road terrain of Anza-Borrego was within a few hours of Los Angeles. However, it was something these typically wealthy Ranger Rover customers from the Los Angeles area would not attempt themselves. The dealer organized the event. The customers paid to stay at La Casa del Zorro and used their own Range Rover for all of the off-road driving. Any damage to the vehicles incurred by the customer was covered by the dealer. This was the second off-road event I had attended and like the first one it was clear the customers not only enjoyed the experience but it helped cement the dealer–customer relationship. It had been another beneficial week or so of offering my support to Range Rover dealers in the USA.

In November 1992 I made a further visit to France to progress the franchise review. This time my visits would be to acquaint me with our dealers in the south of the country. Following a brief meeting with Chris Franklin I flew with J.-M. Mechin from Paris to Marseille where we met up with the Regional Manager Max Bert to visit some dealers. I was not impressed with the standards of the dealers' Land Rover activities, as once again the franchise was mainly an add-on to the main Rover car business.

The next day, November 11th, was a national holiday, Armistice Day, or Armistice de 1918, as it is known in France, marks the anniversary of the end of the First World War. I expected to continue visiting dealers on this day but I didn't. I did a bit of sightseeing in the morning and then met Max Bert for lunch. In the evening Max and I were invited to join the dealer Christian de Boussac and his parents at their home for dinner, which was a very pleasant change from a normal hotel meal. It was after dinner that Christian's father explained the important differences in the world of brandy between a Cognac and an Armagnac and said that the latter was superior to Cognac. We enjoyed some of Christian's father's Armagnac as he wanted to prove the point.

Armagnac is a distinctive brandy produced in the Armagnac region in Gascony, south-west France. It is distilled from wine usually made from a blend of grapes including Baco, Colombard, Folle blanche and Ugni blanc, traditionally using column stills rather than the pot stills used in the production of Cognac. Armagnac is traditionally distilled once, resulting in 52% alcohol. The resulting spirit is then aged in oak barrels before release. This process results apparently in a more fragrant and flavourful spirit than Cognac, where double distillation takes place. So now you know!

On 16th November I moved office to the top floor of Block 1 at Solihull, where other Directors were situated. I had an office next to John Russell's and his secretary Lynne Sperring, who would now provide me with secretarial support. This sadly meant that several weeks earlier I had to tell Kath Duval that her job as my secretary would come to an end. On 19th November I went back to Rover France's offices to help J.-M. Mechin finalize the Land Rover franchise strategy and franchise plan on a location-by-location basis. My work with Rover France was now over for a while.

At the request of Graham Morris, Managing Director of Rover Europe, my next job was to visit Rover Portugal to carry out a comprehensive franchising review and develop a franchise strategy and action plan similar to that in France. In December 1992 I discussed this with David Panton, the Rover Portugal MD, and agreed I would start the process early in 1993.

So 1992 had been my first year where most of my work had been outside the UK. It had involved much travelling by air and many weeks in USA, Australia and France. And as a result I spent many nights away from home. In the seven months until 19th November I was away from home on sixty-eight nights. That was too many!

It was hard to measure or determine the success of my work in 1992 compared to my previous jobs where actual sales results against sales objectives decided success or failure. In many ways I was acting as a business consultant using my knowledge, experience and skills to help and benefit Rover Group teams outside the UK. My new role in 1992 was rewarding but in a much different way to what I had been used to for most of my career up to then.

## Chapter 16
# Rover Group 1993

**IN MY 1993 POCKET DIARY** is an introduction from John Towers, Group Managing Director, Rover Group, which read, 'As you know, our vision is to provide Extraordinary Customer Satisfaction, and even though 1992 has been another very demanding year we have managed to make further progress. Rover Tomorrow is well under way. We are a leaner organization, but we must strive even harder to remove waste, improve our processes and achieve new best practice in everything that we do.'

I thought using the word Tomorrow was unfortunate because 'Tomorrow' never comes – as in the well-known pub promotion 'Free beer tomorrow'. In my opinion to say we should achieve 'new best practice in everything we do' was neither realistic nor achievable from where Rover was at that time compared to its automotive competition.

He went on to say, 'Clearly we will face further challenges in the year to come, and producing the necessary improvements will demand even more effort and commitment from all of us.' To demand more effort was in my opinion the wrong thing to say, because we were all working hard and long hours and as far as I was concerned demanding more effort and commitment implied we were somewhat lacking in these areas, which I don't think was the case. We probably needed to work smarter rather than harder.

John Towers concluded by saying, 'Equally clearly our skills, our dedication and the pride in what we do are assets that, in some areas, have amazed the world at large – and I don't think they've seen the half of it. Thank you for your efforts in 1992. I would like to wish you and your family a safe and happy Christmas, and may 1993 be prosperous for us all.'

At the back of the diary there was a comprehensive summary of Rover Group products and details of Rover Group's mission, which was *'to be internationally renowned for extraordinary customer satisfaction'*. I think that while extraordinary customer satisfaction was a laudable mission, delivering it doesn't mean customers will be loyal to the brand and buy its products again.

In January 1993 one of my first important meetings was with Graham Morris, Rover Europe's Managing Director, and John Russell at Rover Group's Bickenhill Head Office, near Birmingham Airport. The purpose of the 13th January meeting was to discuss Land Rover franchise development, primarily in Europe. I had to prepare a presentation of my 1993 Business Plan and also produce a memo on the subject of Land Rover franchise development for Graham to send to Rover Group's National Sales Companies (NSCs). There was a similar plan to communicate our Land Rover development intentions to the independent European distributors.

Later in January Graham Morris informed me that my Director, Business Development role would now form part of a new Rover Group European Operations function, to be headed by John Parkinson, to whom I would report. My work and activities now would be influenced by Rover Group considerations rather than those from a purely Land Rover viewpoint.

It became clear fairly quickly that the development and growth of the Land Rover business within the Rover Group's European NSCs was essential for their future profitability. In addition, the European NSCs' performance against objectives and overall business activities would now be subject to greater Head Office scrutiny, which would take the form of Quarterly Business Reviews, attended by Graham Morris and his Directors. 'Grow in Europe' would be one of Rover Group's key objectives.

The proposed structure was incomplete and a certain amount of recruiting was underway. However, I recall that a number of potential recruits were not keen to leave their jobs in Rover Cars UK and Land Rover UK to join the new Rover Group European Operations team. Some were persuaded, but others decided they were better off where they were – better the devil you know! A few of these were in Land Rover so their decision to stay there was understandable.

On 16th February 1993 I travelled back to the USA to attend a two-day President's cabinet meeting. The main agenda item for the meeting was the action plan for Land Rover Center development in the USA. Bill Jacobs, the owner and dealer principal of a Land Rover and BMW dealer in Chicago, made a proposal to develop a Land Rover Centers operation covering the Chicago Metropolitan area from a number of strategically located premises. This would be among the first new Land Rover Center of many. Even though I now had a number of European market priorities I was very pleased to continue to support the Land Rover North America team in its major franchise development activity.

In March 1993 I was in Portugal once again to continue my work on the NSCs' franchising plan. It was important for me to visit as many of the existing dealers as possible, not only to understand and see at first hand the existing standards, quality and competence of the dealer network but also to determine the attitude, ability and willingness of the individual dealers to invest in and develop the Land Rover franchise.

In April 1993 I visited Rover Spain, based in Madrid, and headed by Jacques Muller, Managing Director. My two days with Jacques and his managers included a business and franchise review and a series of dealer visits in Madrid, again to get some first-hand knowledge of the existing dealer network. Jacques told me he arrived in Spain from BL's European office in Geneva in the early 1970s to close down Austin production at its factory in Pamplona. However, BL had a sales office in Madrid and Jacques ended up staying in Spain to run BL's sales and distribution and later he became President of BL's Spanish NSC.

Back in UK from Spain I was soon on a flight to Rome to attend my first business review with Rover Italy, headed by Salvatore Pistola, Managing Director. The following week I was on a flight from Birmingham to Paris to attend a business review with Chris Franklin and his management team at Rover France.

On 18th April I flew to Porto, Portugal's second-largest city, to meet Gregorio Luis, the NSC Sales Director. He had arranged some dealer visits in the city, but I hadn't expected to have a memorable dinner there. We were staying at the Sheraton Hotel, but Gregorio took me on a walk down to the port area. Suddenly he stopped and although there was no sign of a restaurant he opened a door and we entered what seemed to be someone's house. We entered a room where there were a small number of tables already occupied by locals. Gregorio said they were probably local fishermen.

Gregorio told me there was no menu and we would eat whatever the owner had cooked, but it would be fish. A lady brought to our table a very large bowl which contained what I would describe as a very thick fish stew. With some bread this comprised our dinner. Everything of the fish was in this large container, fish heads, all the fish bones, as well as some edible fish. It took very careful spooning to avoid the heads and bones and as someone who likes fish it was acceptable. I was told this was an eating place solely for local people, and that accordingly I as a non-local was a very rare visitor to this establishment.

In May 1993 I visited the USA again to participate in the next important stage of the Land Rover Center development. Charlie Hughes arranged for his dealer cabinet members to join him, his senior management team and the senior members of Design Forum at SaddleRidge in Beaver Creek, near the ski resort of Vail in Colorado. SaddleRidge is a premier mountain resort and it was a great place to be locked away from normal automotive business.

The main purpose, over four days at SaddleRidge, was for the dealers to understand the full details of the proposed Land Rover Center concept, including the facilities and standards, and through a series of carefully developed business scenarios to work out for themselves the profitability and return on investment that would result by developing a Land Rover Center in their own location. The programme at SaddleRidge was called 'Find Your Way to the Center'.

The process developed by Charlie and his team was an excellent way for a motor manufacturer to get its dealers committed to invest in its franchise. The dealers were informed they had to make up their own minds on the viability of the Land Rover Center concept, based on creating a financially detailed business plan, and using facts and figures relevant to their own marketing area. At the end of the work sessions the dealers made presentations of their business plans. They all welcomed this 'working together' team approach, carried out within a high-quality resort environment.

On my return to the UK I found myself involved in a number of 'inward-looking' company projects. As part of the Rover Group's focus on 'Extraordinary Customer Satisfaction', a number of subgroups were formed. I joined the 'Manage the Franchise' subgroup.

One of the activities was to develop a 'Best Dealer in Town' programme on the basis that achieving high dealer standards in all locations across Europe would be necessary to deliver Rover Group's vision of 'Extraordinary Customer Satisfaction'. This was a very laudable programme at the time and much work was done to identify the specific standards that would be necessary to be implemented and then measured as the best in town. In practice it was not very realistic or achievable.

There is a reference in my diary dated 12th June 1993 to the opening of the Disney Store in Birmingham, because my daughter Gillian had got a job there. She had told me about the job interview process which I thought was excellent and innovative. I hadn't heard of this method before and even now, if you can do it, I think it's a great way to judge a potential candidate's personality and attitude.

Those applying for a job at the Disney Store were invited to a hotel and were checked in on arrival. They were then given some tests. At the end of this a few were asked to remain while others were asked to leave and were then presumably told they were unsuccessful. The senior executive of Disney then came in the room to tell the few they were successful.

The Disney executive turned out to be the person checking in the job-seekers on arrival at the hotel and as such he was able to make some judgements of these young people in their unguarded moments. Gillian had arrived at the hotel in smart attire and wearing a Mickey Mouse tie, which the Disney executive commented on, and this factor together with Gillian's bubbly personality and their brief conversation made such a positive impression that she had virtually got the job there and then. I have remembered that recruitment incident and process ever since but have never had the opportunity to use it.

Back to my job. The new European Dealer Operations organization would be based in Longbridge, Birmingham, so on 9th August 1993 I moved to a new office there – back to where I started work in September 1969. On 11th August 1993 I played in the Weetabix Women's British Open pro-am golf tournament at Woburn. Our team comprised Rex Holton, dealer principal at Woburn Land Rover, his partner Penny Wells and a young twenty-two-year-old professional golfer called Annika Sörenstam. At the time Annika was just starting her career and was not well known. During our round of golf I was amazed to see how far and how accurately this young lady could hit a golf ball. Annika became the greatest female golfer of her generation, and is regarded by many as the best female golfer of all time. She won ninety international tournaments as a professional, making her the female golfer with the most wins.

At the end of September 1993, I revisited Rover Italy in Rome for business meetings. At the end of the first day I went for dinner with some of the Rover Italy management to a restaurant called Hostaria dell'Orso. I remember we had to sign in before climbing some stairs to the dining area. We had a nice meal, nothing exceptional. At the end of the dinner my Italian colleagues passed the bill to me with the words, 'You said you would pay for dinner next time you came to Rome.' Apparently I had said this on my previous visit but I hadn't remembered. Nevertheless I looked at the bill, which had many 000s in lira in the total. I quickly calculated that the cost was well over £100 per person which was a lot then but I had no option but to pay. It seems this restaurant was an 'in place' to go, and so my Italian friends decided it would be a good place to go for dinner if someone else paid. It would not be the last time I went there.

I had responsibility for fleet sales development in Europe, which brought me into contact with most of the major car rental companies. One of these was Thrifty Europe and we were invited to make a presentation to the Thrifty Europe conference on 15th October 1993 at the Adare Manor Hotel near Limerick. This visit to Ireland also provided me with an opportunity to meet up with Mike Dean, the MD of the Rover Ireland NSC in Dublin, and to discuss business with him. Adare turned out to be a great venue for a conference, with dinner and traditional Irish musical entertainment provided there at the Dunraven Arms.

During October and November 1993, Alan Edis, a Rover Group Director, and I were involved in a project that was considering the possibility of establishing the first joint Rover Group/Honda Vehicle distributorship in Europe. At this time the Honda Motor Company still owned a 20% shareholding in Rover, but vehicle distribution was still run separately by Rover and Honda.

Alan and I flew to Geneva on 23rd November to meet Claude Sage, the President of Honda Autos Suisse, to explore the joint venture possibility in more detail. We had a very productive visit and we left believing it would be possible for the Honda distributor business in Geneva to accommodate and successfully operate the Rover Group distribution for Switzerland.

However, at the end of 1993 I and maybe many other Directors in Rover were not to know that there would be events leading to a major change in the ownership of the Rover Group in early 1994. This would make that first possible joint Rover Group/Honda vehicle distribution plan in Europe impossible to progress.

Chapter 17
# BMW Group 1994–1997

**ON 31ST JANUARY 1994 BRITISH** Aerospace (BAe) announced the sale of its 80% majority share of Rover Group to BMW for £800 million. How had this happened? British Aerospace had bought the Rover Group in 1988 with only a requirement to keep ownership for at least five years. So it was clear even as British Aerospace acquired the company that Rover's ownership could change in 1994.

George Simpson, MD of the Rover Group, travelled to Japan towards the end of January 1994 and met with the directors of Honda, hoping that as British Aerospace was planning to sell its shares, Honda Motor Co. would be willing to make a quick decision to increase its shareholding in Rover and maybe take a majority stake or even 100% ownership. However Honda directors were not willing or able to make a decision over the weekend of 29th January 1994 and in any event I understand Honda top management firmly believed that the Rover Group should remain under British ownership.

On the afternoon of Sunday, 30th January I received a phone call from a Rover Main Board Director telling me very simply that BMW was now Rover's new owner. I imagine Honda directors were more than upset at how the sale of Rover had been handled and it was no surprise to me that on 21st February 1994 Honda announced it was selling its 20% share of Rover Group to BMW. This terminated the alliance of Honda with BL/Rover, which had been in existence since 1980.

BAe had been experiencing financial difficulties in 1991 and many actions were forced through in order to cut costs and/or generate cash.

During its ownership of Rover, BAe had sold parts of Rover's Canley site in Coventry and part of its Cowley site in Oxford – considered by many to be asset-stripping to generate big sums of cash. BMW was looking for ways to grow its sales volume and its directors were attracted by the iconic brands such as Mini and Land Rover. At this time, BMW did not make four-wheel-drive vehicles although one was under development. However, Land Rover was an attractive brand that could be available.

In acquiring the Rover Group, BMW made mistakes. It is generally understood by business experts that there was a lack of proper due diligence owing to the speed of the purchase, and it could have been predicted there would be a clash of cultures between the German and British management that would hinder the successful integration of Rover into BMW. Also some business commentators would later identify poor leadership at the top of BMW in managing the Rover Group business, allowing it for too long to operate fairly independently almost as if there had been no change of ownership.

BMW completed the takeover in only ten days, so may not have looked closely enough at the operation of the businesses within Rover. Had it done so, it may have had a much better view of Rover's problems such as financial losses in the Rover Cars business.

While it was agreed BMW would buy the Land Rover business it was told it would have to take the Rover Cars business as well, although this was thrown into the deal at no extra cost. In practice this 'buy one get one free' purchase should have sent out warning signs to BMW top management as to the very fragile state of Rover Cars.

From 13th to 22nd February 1994, we held a major European region business conference at the Gleneagles Hotel in Scotland. In addition to the 1995 model-year launch of the Range Rover and Discovery I was to deliver a presentation entitled 'Best Dealer in Town'. This was a key operational initiative to raise the quality of Rover Group dealer networks in Europe and my presentation would be the first time Rover's NSCs and independent distributors would hear about the programme and of the action plans they would be required to develop and implement.

My presentation was not delivered from a lectern, but I sat in a chair on a stage and read a script that would be projected on a large screen at the back of the conference room more than twenty yards away. The aim was to appear as if I was speaking informally to the audience. With five events taking place I would make the presentation five times, first to Spain and Austria NSCs, then to France, then Germany and Italy, then independent distributors, then the Netherlands and Switzerland.

I delivered my presentation in English and what I said was simultaneously translated to the audience, most of whom would wear headphones to get the translation. All went well until the third presentation, to Germany and Italy. Fairly frequently the moving script stopped, which meant the autocue process had stopped. This was not a presentation where I could ad lib so I had the discomfort of having to wait each time the autocue stopped until the autocue started again. Fortunately I had a glass of water which I sipped from during the various stops.

At the end of the presentation I wanted to find out what went wrong. I was told simply that the autocue operator had to wait quite often for the person doing the German translation to catch up and finish translating, because German words are usually longer and take longer to say. A rational answer but it was a pity I wasn't warned about it before I started.

The Gleneagles Hotel in Auchterarder, Perthshire, was a great venue for a business conference and I enjoyed staying there. It had been a few years since I was there before, but on arriving at the hotel entrance and on getting out of my Range Rover I was greeted with the words, 'Welcome back, Mr Sparrow.' A very nice way to be greeted indeed!

Gleneagles was also an excellent one to start the full day's ride-and-drive activity of the Range Rover and Discovery, which included a lunch stop at Blair Atholl Castle. At the end of the day, we held a gala dinner at the hotel to complete a memorable conference.

There was no mention of BMW at that conference. Following BMW's acquisition of the Rover Group we were being left alone for a while to carry on running the business as before.

A few weeks later I attended a BMW/Rover liaison meeting at the offices of BMW France in Paris with Chris Franklin, the MD of Rover France. It was my first on-territory meeting with BMW management and the purpose of this liaison meeting was to explore ways of working together while retaining the separate business organization. In attendance at this meeting was a senior executive from BMW's Head Office in Munich who I think was my opposite number. The meeting started at 10 a.m. and was due to end at 4 p.m. We had covered everything we needed to when at exactly 4 p.m. the German BMW executive stood up, clicked his heels, said goodbye to us and left the room. It was my first experience of understanding that meetings with BMW management started and finished on time exactly as planned.

At this time, I visited the European NSCs to discuss their plans for the launch of 38a, the next new-model Range Rover. My meetings included the subject of franchise standards, as we wanted to make sure Land Rover dealers selling the next-generation Range Rover were equipped to meet the overall quality requirements of handling what would be a more upmarket vehicle. The Range Rover P38A as it was known prior to launch was named after the factory building in which the production line was based, but during the development stage the vehicle was also known within the company by its project designation of 'Pegasus'.

The 38a would be launched to the public across Europe on 28th September 1994, twenty-four years after the introduction of the first-generation Range Rover. The new model would include an updated version of the Rover V8 engine, with the option of a 2.5-litre BMW six-cylinder turbo-diesel. The new model would offer more equipment and premium trims, positioning the vehicle firmly in the luxury-car sector and further above the Land Rover Discovery.

In June 1994 I attended a major Range Rover dealer/customer off-road event held in the Anza-Borrego Desert State Park located within the Colorado Desert in southern California. I had previously supported a similar Range Rover customer event and once again my presence as a Land Rover Head Office Director was greatly appreciated by Dick Bauer, his Bauer Anaheim dealer staff and their customers. I can remember it was exceptionally hot in the desert, and on one day my diary shows we had a weather temperature of 112 degrees Fahrenheit or 44.4 degrees Celsius. We called it fire heat.

I added some holiday to the trip and my wife Annette and I spent some days at Laguna Beach and Santa Monica in the Los Angeles area. Staying in Santa Monica enabled us to do some sightseeing in Hollywood and some shopping in Rodeo Drive, in Beverly Hills.

In July I attended a meeting at BMW's Head Office in Munich. It was the first meeting where my department's managers and I would present our business plans to our opposite management in BMW. The BMW management comprised Dieter Kirchmeier, Michael Hruska, Jurgen Tegkmaier, Peter Gesche and Arthur Dalziel. Arthur had previously been an Area Sales Supervisor in Austin Morris at the same time as me, and was now BMW's Corporate Sales Manager.

There was an interesting start to our meeting in one of the conference rooms. I sat in the middle of one side of the table with my managers, whom I think included David Carpenter, John Stretton and Barry Tanser either side of me. Dieter said I should sit at the head of the table as he said he was expecting me as the head of the department to make the full presentation of the Rover team's business plans. He was surprised when I said each of my managers would present their individual plans, including franchise standards, dealer development and corporate sales, and that we would sit where we were. This was another small but significant example of the different work culture in the two companies. I wasn't going to change the way either I or we in Rover Group operated just because my German opposite number said so.

Our discussions continued into a second day. It became clear that unless instructed otherwise our business activities would continue relatively unchanged as there was no obvious way to integrate the two separate company plans. However, it was agreed there was an opportunity for Rover and BMW to work together in growing vehicle sales to the corporate business sector in Europe and internationally. BMW was already strong in this sector; Rover was not. John Stretton and Arthur Dalziel were charged with putting together a corporate sales strategy, including a common procedure and action plans to exploit this opportunity.

In September, on returning from a two-week family holiday to Nerja in the south of Spain, I was informed that I would be travelling two days later to Gibraltar to support A. M. Capurro the Land Rover distributor in launching the new-generation Range Rover (P38A) as mentioned earlier.

Gibraltar is a British Overseas Territory at the southern end of the Iberian Peninsula and at the entrance of the Mediterranean. The famous Rock of Gibraltar is the only landmark, and at its foot is the densely populated city area, home to almost 30,000 Gibraltarians, who depend heavily on Spain for produce and supplies. English and Spanish are widely spoken and thousands of Spaniards cross the border each day to work. Spain ceded Gibraltar's sovereignty to Britain in a 1713 treaty but has persistently sought its return ever since. I had not been to Gibraltar before so I was looking forward to my visit.

I was met at Gibraltar airport by John Capurro, the owner of the business, and we set off the short distance to the Capurro premises in Line Wall Road in an elderly Austin 1300 saloon. What made the vehicle rather special was the union flag mounted royally on the front of the car's bonnet. It soon became very clear to me that the people of Gibraltar were proudly British and the wish to stay so by almost the whole population continues to this day.

We had some work to do to get the Capurro management ready for the evening launch of the Range Rover. I had prepared a speech which covered the history and development of the new vehicle, the importance of Capurro, the Land Rover importer in Gibraltar, and I referred to our aim to exceed Range Rover and Land Rover customers' expectations with exceptional customer service.

That evening John Capurro and I launched the new Range Rover to 150 invited customers and guests. I remember speaking to a Mr Gilbey, who told me he was from the Gilbey's Gin family, was involved in the sherry business, and was living in Jerez, in the south of Spain. He had travelled especially to Gibraltar for the Range Rover launch. I also had a conversation with Gibraltar's Chief of Police. There was much drink flowing that evening and I asked the Chief of Police how they handled drunken behaviour. He said that one possible action was to ask the person under suspicion of drunkenness to walk in a straight line. I don't know if it was a serious answer to my question. I judged the evening and the new Range Rover launch to be very successful and enjoyable for all concerned. It was certainly one of the more enjoyable vehicle launches I have been involved in.

The following day we had a tour of Capurro's premises. What I remember most is going to an underground level where there were garage-like enclosures, which when the doors were opened revealed old but brand-new cars such as Triumph sportscars. It seems there was a strange wish within the Capurro family to 'salt away' many historically valuable cars for posterity. In the afternoon I flew back to Gatwick.

In the last half of 1994 I took part in the PR3 Steering Group meetings. The idea that the Rover Group should develop a small two-seater MG sportscar existed through the 1980s but was not a priority given the company's financial difficulties during this period. However, when the Mazda MX-5 small two-seater sportscar was launched in 1989 to great success it allegedly provided the necessary encouragement for Rover's top management, now under BAe ownership, to decide to enter the same market. As a result in the early 1990s the PR3 project that was to develop and introduce the MGF sportscar progressed.

Now on one December afternoon and night at the end of 1994 the PR3 Steering Group met at Gaydon in Warwickshire to carry out a driving assessment of a base model and higher-specification model MGF alongside a Mazda MX-5 and a Toyota MR2. I remember the briefing for the driving session because it was raining and we were told not to crash either of the MGFs because they were the only ones that existed. The MGF was launched on 8th March 1995, and it made a positive impression on both the public and the press. It was the first virtually all-British car produced by Rover for a decade.

September 1994 represented a career milestone for me. At the start of the month I completed twenty-five years' continuous employment albeit the company ownership had changed a few times during that period. On 9th December 1994 a long-service dinner was held at the National Motorcycle Museum near Solihull for all those persons completing twenty-five or more years' service with the company. That night I collected my suitably engraved long-service pocket watch from Graham Morris.

One of the first indications of organizational change following BMW's acquisition of the Rover Group came when Graham Morris announced he was leaving. His 'leaving do' took place on 16th March 1995 at Studley and was attended by Rover Group Directors and senior management. I cannot reveal what Graham was wearing but let's just say the analogy he used to describe his leaving referred to a Second World War escape. It was typical of Graham to treat his departure from Rover with some entertainment and there was a high degree of hilarity for those of us that were there.

The original Rover 400 was developed in collaboration with Honda and introduced in 1990 as a series of small family cars. Now in April 1995 I attended the second-generation Rover 400 launch in Madrid. The car was launched as a hatchback and later as a saloon, while Honda's equivalent Civic was just sold as a hatchback in the UK. In my opinion it was an OK car, but not very exciting.

At the end of April I represented the Rover Group at the funeral in Zurich of Mr Emil Frey, the Rover distributor in Switzerland. Emil Frey had created in 1924 what was to become a large car-importing company for many brands including Rover and he had now died at the age of ninety-six. He had been selling British cars since 1931 so it was fitting that Rover should be represented. Richard Wright, a Jaguar Cars Director, represented his company.

In May I had my first ever computer training. I received a large desk computer and it was now time to learn how it worked and how to use it. I hadn't asked for a computer so someone must have decided I needed to have one. I know I used it initially to access company information rather than as a means of communication. In 1995 I was still sending paper memos, faxes and letters both internally and externally, and being helped by a secretary in the process.

In February I received an invitation from Arthur Dalziel to play in the BMW International Open pro-am golf tournament in Munich. I would be joining a group of UK-based BMW Corporate customers as well as some BMW GB executives to Munich, most of whom would be playing in the pro-am event.

On the evening of 27th June in the Hilton Park Hotel we had dinner and after that gathered in a conference room to find out which professional golfer would be in our team. There was a big board displaying the tee times and alongside these were the names of the professional golfers. These included such well-known golfers as Darren Clarke, Bernhard Langer, Sandy Lyle and Lee Westwood. The names of the amateurs were then drawn and matched against a professional golfer. I saw that one of the amateurs playing was the legendary German footballer Franz Beckenbauer.

There were thirty-nine teams with the first team teeing off at 7:30 a.m. and the last team teeing off at 2:10 p.m. We were all wondering which professional we would get and hoping for decent tee-off time. The result of this procedure was that I was to play with David Voss, Director of

VELO, and Keith Thomson of Pitney Bowes. The professional golfer in our team was Anders Forsbrand of Sweden. Our tee-off time was a nice 11:10 a.m. at the St Eurach Land und Golfclub, which was some 25 miles south of Munich. So fortunately we didn't have to get up very early.

The round of golf would be determined by the best two scores of the four players in each team. We played well and on almost every hole we had two very good scores. The weather was good and we were enjoying the friendly atmosphere. Anders was supportive and said he would not comment on our golf swings and as it was his practice round we were aware of the need to be professional in what we did and how we conducted ourselves. Even if you are an amateur your behaviour and how you conduct yourself on the course is very important. Halfway round we stood together to have our photograph taken.

On the 11th hole (our second hole), a 368-metre par four, I got a net eagle. However, the 2nd hole (our eleventh hole) has to be my most memorable two golf shots on one hole. The hole was a 165-metre par three over a lot of water. I took a 7-wood and hit a good shot which ended pin-high but some 20 feet to the right of the hole. The green sloped steeply from back to front so my putt would have to be aimed up the green with the hope it would descend in a curve towards the hole.

For the first time during the round of golf I asked Anders for some advice as we both surveyed the direction of the putt. After a short while he said a few words I'll remember forever. 'Hit it hard but not too hard.' I visualized the direction I wanted the golf ball to take, didn't wait too long over the ball and hit it. The ball went up and round and down the slope in an arc and to my amazement it went straight into the hole. Anders spoke just one word to me: 'Superbo'. Given the fact we were in a serious match, and that it was a very difficult putt that could easily have gone very wrong, it was in the circumstances definitely one of my best ever putts.

At no time did we know as a team how we were doing. There were no boards to show team scores and I was only keeping a record of my scores. Even when we finished none of us knew how well we had done. Anders went off and we went to the changing room. We then went to get some drinks and chatted to other members of the British group. In due course the coach took us back to the Hilton Park Hotel where we got changed for dinner and continued some drinking.

That evening there was a dinner in the hotel for all the guests and amateur golfers. There were no professional golfers present. I was on a table with David Voss, Keith Thomson, a number of BMW's British Corporate guests and John Stretton, my Corporate Sales Manager. We had an enjoyable dinner with a good deal of alcohol and some inevitable speculation about who had won. I was told we wouldn't know until Dr Wolfgang Reitzle, who held the number-two position in BMW, announced the results and the winning professional golfer appeared.

In announcing the results Dr Reitzle spoke in German and read out the names of those in third place, who had a net score of 59, or 13 under par. I thought we had done better than that. Dr Reitzle then read out three German names, who had a score of 57, or 15 under par. It was only then that Anders Forsbrand appeared and I realized we had won. Dr Reitzle read out my name, David Voss's, Keith Thomson's and Anders Forsbrand's and announced we had won with a score of 56, or 16 under par. There was great cheering from the relatively few British present as we, the winning pro-am team, walked up to the stage to be congratulated by Dr Reitzle.

We were each presented with a PING tournament golf bag, some PING metal woods and a BMW-branded silver cup. Our prize also included a holiday at a Hilton hotel. David and I politely said that as we were left-handed golfers we would be grateful if BMW could replace the right-handed clubs with left-handed versions. Anders, on behalf of the team, did the usual winner's speech of thanking everyone, especially BMW for its sponsorship. Some photographs were then taken of our winning team. It was good that we had a photograph of our team taken while we were out on the course because despite asking several times I never did receive a copy of the presentation photograph taken with our prizes and cups.

Much later Dr Reitzle came to our table to say well done to us again.

As we stood up he pointed a finger to my chest and said sternly, 'John, next year a German team will win the pro-am.' He then added that we, as the winning team, would be invited back the following year to defend our title, an invitation for which we thanked him.

During the months of July to September I had a number of meetings with BMW management at their Head Office in Munich. On arriving at the reception desk at the main entrance of the Four Cylinder building I and other visitors had to leave our passports with the receptionist before being allowed to go further. After a number of visits I was recognized by the receptionist but still I had to comply with the passport process. Security was taken extremely seriously by BMW at its Head Office.

What I found interesting at the time of my visits to BMW Head Office was that all the taxis from and to Munich Airport were cream-coloured Mercedes-Benz cars. Several times I asked the taxi driver to explain why this was the case in Munich, the home of BMW. And each time the taxi driver pointed to the dial in front of him that invariably showed the car had done over 200,000 or 300,000 kilometres. They were almost telling me BMW cars were not capable of doing what a Mercedes-Benz could.

During my work with the Rover Group NSCs, the development of Land Rover Centres was an ongoing objective. Until now we really only had successful Land Rover Centre examples in the UK to show the European NSC top management. However, much progress had been made on Land Rover Center development in the USA. So I arranged to take most of the NSC Managing Directors to the USA in October 1995 to see for themselves what had been achieved.

We visited Land Rover Centers at Massapequa and Glencove owned and run by Brian and Mike Lazarus. We also visited their Saturn dealership. We visited a Harley-Davidson dealer in Hempstead. Harley-Davidson was viewed as a premium brand that could provide some learning, especially in the area of accessory sales and marketing. We were told the revenue and profit from merchandise and accessories matched that of the motorcycles themselves.

The following day we travelled south by coach to Prestige Motors of Paramus, New Jersey, a Land Rover and Mercedes-Benz dealer. From there to Philadelphia to see Mainline Land Rover, and then to the Land Rover dealer in Wilmington, Delaware.

We flew back to London Heathrow from Washington Dulles Airport. It may have been a relatively short trip to the USA but it gave the NSC management some direct experience, knowledge and understanding of the Land Rover Centre concept and its development issues in a major market outside the UK, which at the time was important.

On 23rd November, my wife Annette and I took a holiday to India, one that turned out to be memorable for many reasons. The holiday was scheduled to comprise two nights in Goa, on the west coast, followed by a flight to Jaipur, and then a coach tour to Agra, and then on to Delhi. We were in a guided tour party of about ten couples, and the coach tour itinerary would include all the famous sights, including the Taj Mahal. At the end we would fly back to Goa for another seven nights at our hotel by the beach.

We experienced disruption to our holiday resulting from internal flights being unexpectedly terminated, unnecessary taxi journeys taken and repeated in Mumbai and so called five-star hotels that were nowhere near that standard. On the way to Agra we experienced the sort of driving on the road that was hair-raising. Coaches and trucks overtaking in places that looked like accidents waiting to happen. Occasionally we stopped and were met by small children wanting pens or pencils or anything else we could give them. We were travelling on a main road where we could see poverty all along. It was a real eye-opener and our travel was becoming more of an adventure than a holiday.

Later that afternoon we visited in the city of Agra and one of the most famous buildings in the world, the Taj Mahal. The Taj Mahal is an ivory-white marble mausoleum on the south bank of the River Yamuna. It was commissioned in 1632 by the Mughal emperor Shah Jahan to house the tomb of his favourite wife, Mumtaz Mahala, who died giving birth to their fourteenth child. It's a stunning building. After that we visited the Agra Fort, from where we could look back at the Taj Mahal. That night we stayed at the Novotel in Agra. I remember that evening well because Annette had become quite ill as the result of being bitten and infected by mosquitoes. On returning to Goa we had time to recover from our coach travels to Jaipur, Agra and Delhi. It was a memorable holiday!

At the start of 1996 I received a Rover Group 'Learning Diary'. In his introduction John Towers, the Rover Group Chief Executive, stated: 'The quality of our people, products and processes is what will define us as a "Class Act".' However it was very optimistic because I didn't see the Rover Company realistically ever becoming a class act. In 2000 BMW sold Rover and certain MG activities to the Phoenix Consortium which became insolvent in April 2005.

During 1996 I made a presentation to all HR staff working in the Rover Group. At the Q&A session at the end I received a question and asked the person's name. He said Fred Herlihy. I saw him afterwards and said, 'Fred, I am pleased you kept an eye on me.' He looked blankly at me but these were his words to me some twenty-seven years earlier on my first day at work.

In the first week of January I was contacted by John Towers with the question, 'What are you doing next week?' Even if I had a full diary, which I hadn't, I was obviously going to be doing something I hadn't planned. On 8th January 1996 I was in a senior management meeting discussing a project on Bulgaria.

Towards the end of the Maestro production, Rover top management decided on a plan for the Maestro to be assembled in Bulgaria, with tooling and assembly line sent from the UK. The cars would be built from CKD (Complete Knock Down) kits in a joint venture with the Bulgarian government at Varna on the Black Sea coast.

I was told Rover and the Bulgarian government had high hopes for the Maestro project, and that there was a target to build up to 10,000 cars a year. The investment in the plan was estimated at $20 million, but I learned that things were not going to plan. Cars were being built but there were very few customers for the finished vehicles. As a result I was asked to go immediately to Bulgaria, carry out a full assessment of the existing sales situation and future sales prospects, and then produce a report with recommendations.

On 9th January I flew British Airways from Heathrow to Vienna and then on Bulgarian Airlines from Vienna to Sofia, the capital of Bulgaria. At Sofia Airport there was a queue at passport control where I was surprised to be asked for cash to go through. I had been warned about this possibility and so I handed over some US dollars.

In the arrivals hall I was met by Allen Pengelly and Andy Myers, senior Rover managers who were working on this Maestro production venture. While in Sofia I met with the senior management of Daru Car. Daru Car was founded in 1992 and started its business as an official importer of BMW for Bulgaria. In late 1994 owing to the purchase of the Rover Group by BMW, Daru Car became the official importer of Rover and Land Rover for Bulgaria. The Daru Car management were very cooperative and helpful in providing me with information, data and views on the Maestro production and sales situation.

All this enabled me to get a very good understanding of the business situation for the Maestro. Over the days I was there it became clear to me that Maestro production in Varna could not continue, because there was almost no demand or customers for the vehicle, and no prospects that this situation would improve.

Since starting their assembly in late 1995 most of the Maestros that were built were now being parked and stocked in their hundreds. I was told that the majority of the Maestros should have become vehicles for Bulgarian government officials and management, but this wasn't happening.

Another motor industry contact with knowledge of these events has said, 'There was the noncommitment by the Bulgarian Government to honour substantial, previously promised, departmental orders for thousands of the vehicles – a measure that would have raised the profile of the Maestro in Bulgaria.'

I left Bulgaria certain about what my report would recommend. I wrote my report which simply said there was no business opportunity for the Maestro venture in Bulgaria and that Rover should exit the venture as soon as possible. I can recall hearing from John Towers that my recommended action was not what he wanted to hear. On 4th April 1996 production of the Maestro in Varna stopped. The factory was closed, which that meant 250 workers had lost their jobs after building 2,200 cars.

Back in the UK we started to investigate finding buyers for the unsold left-hand-drive Maestro vehicles from the fleet and corporate sector in Europe, including the major car-rental companies. However many of the unsold vehicles were brought back to the UK, converted to right-hand-drive and sold cheaply. There is much more than could be written about this Maestro project but this should suffice.

On 20th January 1996 I flew to Tokyo, Japan with John McHugh and David Carpenter to carry out business development and franchising reviews with Rover Japan top management. This was a new experience for the three of us. We were able to visit not only Rover dealers in the Tokyo area but also the very large vehicle showrooms of the main Japanese manufacturers Toyota, Honda and Nissan, all of which had the latest customer experience facilities.

In early February 1996 I was very involved in another country project. This time it was Belgium. It was decided that the Rover National Sales Company in Belgium should take over the vehicle distribution of Land Rover which at that time was handled by the independent importer Beherman-Demoen. There was a great deal to be done to action this significant change and to ensure as far as possible a very orderly transition of the business.

At the first meeting with Philippe Mertens, the Rover NSC Managing Director, we had to decide on an appropriate name for the

Land Rover project. This involved a fairly short discussion of any famous Belgian names we could use and, having ruled out the cyclist Eddie Merckx, the five-time winner of the Tour de France, and the detective Hercule Poirot, we agreed on Valentine because 14th February was Philippe Mertens's birthday rather than because there was any implied massacre in our planned actions. This project would continue for several months up to July 1996.

Later, on 19th February, I set off on a flight from London to Australia. Alan Curtis and Alex Stephenson, Rover Group Directors, were on the same flight. The main purpose of the visit was to take part in a major Land Rover conference at Uluru, then called Ayers Rock, a large sandstone rock formation in the southern part of the Northern Territory in central Australia. It lies 208 miles south-west of the nearest large town, Alice Springs. This conference would not only cover action plans for Land Rover Centres but also Land Rover product development and planned introductions. In addition, and after the conference, I would visit a number of the larger dealers in the big cities of the country with John Parkinson, who was travelling on a different flight. On my first trip to Australia in 1992 I had said I would aim to visit many of the dealers again one day and so this would be my opportunity to do so.

The BA009 flight to Sydney was due to take off at 8:30 p.m. but several times the 747 went to the runway only a short while later to return to the stand. It was cold and snowing and I think each time we returned to the stand the wings were de-iced. I was on the upper deck in club class and could hear other planes taking off but there was no communication on our plane about our delay. Eventually according to my diary we took off at 1:30 a.m., five hours later than we should have done. After we had taken off the captain came on to say that the reason for the delay was because the plane was fully loaded, and in the weather conditions he could not have stopped the plane at the end of the runway if he had to abort a take-off.

The plane had to make a scheduled stop in Bangkok and with the five-hour delay we didn't arrive in Sydney until 10 a.m. on Wednesday, 21st February. As a result we missed our connecting flight to Ayers Rock. We were told we could not now get there until the following day and there was some danger that Alan, Alex and I would miss the start of the conference. Various flight options were being considered to get us there in time, but we had an unplanned overnight stay in Sydney. In the end we arrived in Ayers Rock via a flight from Sydney to Alice Springs, just in time for the presentations. After we joined a Range Rover ride-and-drive event with the Australian dealers which included a tour towards Mount Conner, a flat-topped mountain not unlike Ayers Rock.

The following day was a continuation of the business and product conference at the Sails in the Desert resort at Ayers Rock, where we were all staying for three nights. That evening we enjoyed a gala dinner outside in the desert. At one point after the dinner the lighting was turned off and we were told to look up at the sky. The sight of a sky full of thousands of stars was amazing.

The next day was Saturday, 24th February and at 7:45 a.m., sat alongside the pilot, I took a helicopter ride lasting half an hour around and over Ayers Rock and the Olgas. Kata Tjuta/Mount Olga form the two major landmarks within the Uluru-Kata Tjuta National Park considered sacred to the Aboriginal people of Australia. In the afternoon Brian Smith and Bruce Keown of Bruce Lynton Land Rover, Southport on the Gold Coast persuaded John Parkinson and me to drive out to the Olgas, also known as Kata Tjuta, a group of large, domed rock formations located about 16 miles to the west of Uluru.

On arriving at a visitor parking area, Brian and Bruce suggested we start walking around the Olgas and initially although it was very hot this was fine. As we walked further the air was full of flies. Fortunately I had a baseball-style cap with netting that kept the flies off my face. However, what was bothering me more was that while we were not exactly lost Brian and Bruce decided we should keep walking round the Olgas rather than turn back the way we came. We had by now consumed all the water we had and we had no way of knowing how far we had to keep walking to reach our vehicle.

According to my diary entry for that day we walked for around three hours and probably 6 miles before we came to a very large water tank that had the words 'For Emergency Only' painted on the side. By the time we found this water we were not far from it being an emergency. Shortly after finding water we found our way back to our vehicle. Here was a sign saying no one should go walking around the Olgas without a map, plenty of water and letting others know where you were. I think we had failed on all three counts.

On one of the Uluru websites it says there are two walks open to the public with the following advice. 'If you want to walk around the entire base of Uluru, allow 3.5 hours for the 10.6 km loop. Take plenty of water, a hat and be prepared for lots of flies.' I suppose we did this longer walk, which I shall not forget!

A day later John Parkinson and I flew to Cairns and began a ten-day programme of Land Rover dealer visits, most of whom I was seeing for a second time. In my diary for 2nd March 1996 I entered the following words – 'JMP ref Korea'. While in Sydney John Parkinson asked me an interesting question. Would I like to become the BMW Group President of South Korea? It was a question out of the blue and not one I had ever expected to get. My immediate reply was to say no, to which he replied, 'I didn't think you would say yes.' Had I been a lot younger and without a family it might have been a different situation, but I knew South Korea wasn't for me.

Until now the senior management structure of Rover Group had remained largely unchanged in the two years following BMW's acquisition of the company. Graham Morris, MD of Rover Europe, had left. However, on 29th April 1996 there were a number of changes. It was announced that John Towers, Chief Executive of the Rover Group, was leaving and that John Russell, Sales and Marketing Director, was moving, although the announcement didn't say where to or what to. I was now more aware of the fact BMW was increasingly concerned at the Rover Cars losses and that there would be more changes to come.

On the evening of 13th May I flew to Johannesburg, South Africa, not only to hold a Land Rover business review, but also to discuss the management's plans to create a Land Rover Experience Centre in the country. I arrived in the early morning of 14th May and was picked up at the airport by a driver in a Land Rover. I sat in the front passenger seat a noticed a big gun in between me and the driver. He explained there had been a lot of hijackings at that time in the Johannesburg area and that having a gun in the vehicle was necessary for safety and security. He also told me if he did not stop at traffic lights or suddenly drove off-road not to worry.

I found out later that a number of car owners, many of them in BMW cars, had been ambushed while stationary at traffic lights and were then shot and killed and their car stolen. I also found out that Phil Popham, who had joined Land Rover South Africa as Marketing Director, was living in a secure executive housing estate that was enclosed with high barbed-wire fencing, and the security included guard dogs. I was booked into the Karos Indaba Hotel in Midrand, just north of Sandton, a wealthy district of Johannesburg, for my three nights and was told for my safety not to leave the hotel under any circumstances.

On the first afternoon I was given a briefing on the Land Rover business in South Africa by Piet Rademeyer, the Managing Director, and Phil Popham, as well as on their plans for a Land Rover Experience Centre. As the name implies this Centre would give all customers the opportunity to drive the vehicle in off-road conditions and learn how to do so.

To explore possible locations for the Centre, the local management arranged for a helicopter to take us to several the next day. Our journey included a possible site at Kyalami in Midrand, famous for the motor racing circuit. In recent years, the area surrounding the circuit has developed into a residential and commercial suburb of Johannesburg. From the helicopter we could see there was a very large township close to the possible Kyalami site.

The following day we visited the Land Rover assembly plant at Rosslyn, a northern suburb of Pretoria, and the location also of the BMW assembly plant. The latter opened in 1968, and was BMW's first factory outside Europe. Later that day we had another meeting to discuss action plans for the Experience Centre, which included my suggestions on introducing the concept. On 18th May 1996 after another overnight flight I arrived back in the UK.

At that time BMW only expected its management to speak German if they were based at the company's Head Office in Munich. However it seemed reasonable for those of us based in the UK to have some basic German language competence and so at the end of May 1996 a number of us started a series of German language lessons, each lasting four hours. We had five or six lessons, just enough to give us a flavour of the language.

On 18th June 1996 I flew to Munich to take part in the BMW International pro-am golf tournament in Munich, having been invited by Dr Wolfgang Reitzle of BMW to defend my 1995 title. No luck this time.

After a few years of BMW ownership there had been little organizational change in the Rover Group. However, on 30th July 1996 we did learn of some change. The main change was that Tom Purves, who had been Managing Director of BMW GB, was to take over from John Russell as Sales and Marketing Director of the Rover Group. At the time that announcement was a surprise to me and many others. However, looking back I think it was inevitable given Rover's continuingly poor financial situation that at some point BMW would put some of its top managers into the Rover Group.

However even this didn't resolve the financial problems. In the five years since BMW took over Rover Group it had invested £2.5 billion but the Rover Group's losses in 1998 were around £600 million, mainly because of the strength of the pound and Rover's inferior productivity.

During the few years of BMW ownership I experienced what I regarded as the worst side of senior BMW management, a dictatorial management style, not far removed from the stereotypical German army officer culture of 'follow all orders'. It had been decided by top BMW management that there should be some integration of BMW and Rover Group vehicle distribution in those European Markets where BMW and Rover Group had independent importers. I was to be involved in this.

I remember saying I suggested we carry out a thorough business audit of the importers and select the one best able to handle the BMW and Rover Group product ranges. I was told this would not be the approach. All the brands in the BMW group would be given automatically to the existing BMW importer. While I fully understood why BMW took the position it did, I couldn't agree this was the best or only way to decide the best vehicle distribution arrangements for the BMW Group where BMW and Rover Group had independent importers.

The first market I was involved in was Cyprus, where Char. Pilakoutas Ltd was appointed as the Rover Group distributor (Rover Cars, Land Rover and MG) in 1995. This was not the only market where this 'BMW first' policy was applied.

On 5th October 1996 Annette and I flew from Gatwick to Phoenix to start a two-week holiday in Arizona, USA. We travelled northwards from Phoenix to Sedona, then on to Flagstaff on Route 66. From there we stopped at Meteor Crater, the world's best-preserved meteorite impact site, located 18 miles west of Winslow. The crater is nearly a mile across, 2.4 miles in circumference and more than 550 feet deep. An incredible sight.

We drove through Monument Valley, a desert region on the Arizona–Utah border best known for the towering sandstone buttes. Monument Valley has been featured in many films since the 1930s. The film director John Ford used the location for a number of his best-known Western films starring John Wayne, including *Stagecoach* (1939) and *The Searchers* (1956).

We stayed on the night of 9th October at Gouldings Lodge in Monument Valley. Ford and John Wayne stayed here while making many of their films. The views we had from our room window across Monument Valley were spectacular. The next day we headed to the south rim of the Grand Canyon. The Grand Canyon is too big to describe the views. It is 277 miles long, up to 18 miles wide and has a depth of over a mile (6,093 feet).

We stayed four nights in Tucson. On 15th October we travelled to Tombstone, best known as the site of the gunfight at the OK Corral, but the actual gunfight was on Fremont Street, a block or two away. Tom and Frank McLaury, and Billy Clanton, who were killed in the OK Corral shootout, are buried in the town's Boothill Graveyard. When we visited Boothill I was surprised to see the graveyard was on relatively flat ground and not situated on a hill, as the name implies. Another 'graveyard' we saw in Tucson was the Aerospace Maintenance and Regeneration Center, where some 4,000 aircraft are stored. Some are used for parts, returned to service, or sold to foreign governments, but for many it is their final resting place.

Back home on 22nd October my father was seriously ill so I went to see him. Unfortunately on 26th October he passed away. His funeral took place at St Helen's Church Wheathampstead on 1st November 1996. They say things come in threes. Another visit to a graveyard.

Looking back most of 1996 was a year when my career in BMW was not clear, especially having been sounded out in March about the BMW Group President's job in South Korea. I don't know whether saying no to that job ended my career but if BMW senior management in Munich HQ was involved it certainly didn't help.

Many discussions took place and finally I agreed with David Bower, Personnel Director and Main Board Director of the Rover Group, to retire early on 1st October 1997, just over twenty-eight years since joining BL from university in September 1969. In more ways than one it was a sad ending to the year.

Chapter 18

# Automotive Consultancy 1998–2000

**IN LEAVING MY MOTOR MANUFACTURER EMPLOYMENT** my initial and maybe natural reaction was that I didn't want to seek employment in another motor manufacturer. I felt I wanted to be more in control of what I did in the remaining part of my working life which could be ten years or more. I wanted to take time to decide what to do next. I explored a number of opportunities in Europe with American, German, Japanese and Korean motor manufacturers but none had a job that I was happy to take.

In November 1997 I had an exploratory meeting with Gordon Roscoe and Dean Hines, Directors of Jackson Consultancy Group plc (JCG). At the time it was one of the largest consultancies specializing in the UK automotive industry. I hadn't heard of JCG when we met but as I hadn't been working in the UK since 1992 that was understandable. In January 1998 I had a couple of meetings with Michael Jackson (Chairman) and Rob Purfield (Managing Director) at the JCG Head Office in Bingley, Yorkshire, where they learned about me and I learned about them and JCG.

Michael and Rob were keen for me to join JCG to help develop their business in two ways. First, to use my considerable number of UK automotive contacts and relationships with the top management of UK-based car manufacturers/importers, UK dealer groups and dealers to set up meetings to promote JCG business, consultancy and training programmes. Rob said to me, 'You get us the meetings and we'll get the business.' Second, to use my international knowledge experience and contacts to gain and develop business for JCG outside the UK. I was offered a JCG board position as International Director.

I said I would think about it and let them know as soon as I had decided. I was not sure whether my management style and way of working would fit well with their company's way of operating and company culture.

After much careful thought I decided I would accept the offer JCG had made, knowing that this new job would present a different challenge to any I had experienced before. I attended my first Directors' meeting on 3rd February and JCG's tenth-anniversary gathering a few days later. It was agreed that my short-term focus and priority would be in setting up business meetings with my UK motor industry contacts, meetings which I would attend with Rob Purfield. To aid our 'selling' process a JCG brochure was quickly produced setting out in detail the full range of JCG's business programmes.

Initially I worked from home but I agreed with Michael and Rob that to promote and market JCG in a professional way, especially outside the UK, it was necessary 'to look and act the part' of an international business and to do so we needed to have a company office, not too far from where I lived. We rented a fully serviced Regus office in the Waterside Centre in the Birmingham Business Park and I moved in on 14th May 1998.

According to my diary during the period to end-July I arranged meetings with most motor manufacturers and dealer groups based in the UK. I also considered how we might operate in Europe. I had many automotive industry contacts in France, and a good knowledge and experience of this country, so we could 'put our toe in the water' here. However, to have any chance of obtaining business we needed to employ one consultant who was a French national, but also fluent in English. Ideally it should be someone I knew.

Didier Deboos had been a business consultant in France with Alison Associates, a business management consultancy that worked with the Rover Group. I had a first meeting with Didier at my Birmingham office on 9th July 1998. He was interested and soon after he joined JCG. Even then I did wonder what one person could reasonably achieve. A first step was to translate the JCG business brochure into French.

Honda UK was an important client for JCG and on 27th October 1998 Ken Keir, Managing Director of Honda UK, visited JCG for a business review which lasted most of the day. I had known Ken since the early 1970s when we were both working for BL and so I was invited to join the meeting. At the end of the meeting Ken and I had a few words and we agreed to keep in touch.

Didier arranged a meeting with Peugeot France. We developed a bespoke used-car programme for Peugeot that could raise the standard of used-car activity in the Peugeot dealer network. But no business resulted from our efforts. What I learned from this exercise was that it is very easy for a potential client to request a business proposal from a consultancy and then use the format and details provided by the consultancy, in this case JCG, to develop its own company programme. Unfortunately the proposals we developed and presented had the risk of having our work hijacked by the potential client. This is what happened with Peugeot. That is a business and consultancy lesson I have carried with me from that time.

On 15th December 1998 we had our first meeting to explore possible business in the Middle East. This meeting in our Birmingham office was with Chris Preston, Managing Director of Trading Enterprises of Dubai, importers of Chrysler Jeep and Volvo vehicles in the United Arab Emirates (UAE) and part of the Al-Futtaim Group of companies. The Al-Futtaim Group was a significant 'conglomerate' even in the late 1990s with eight divisions comprising automotive, electronics, insurance, services, real estate, retail, industries and overseas.

It was clear from this meeting with Chris that there were opportunities in the UAE for JCG to deliver professional automotive training courses to vehicle importers, providing the UK-based JCG took the decision to locate and establish its consultancy activities in the market. We learned that UAE companies did not want what they called 'fly by night' consultancies that flew in their consultants to deliver some training and then flew them out again. All JCG Directors agreed we should explore these opportunities further. Gordon Roscoe, who headed JCG's training business, and I made visits including the UAE in 1999.

Despite our best efforts no business was forthcoming in France and because his income was largely dependent on gaining business Didier Deboos left JCG in April 1999. It was clear that getting automotive consultancy business outside the UK was going to be exceptionally difficult.

As all JCG Directors had agreed we should explore the potential business opportunities in the Middle East it was down to me to set up meetings with possible clients. Fortunately I had a number of good Land Rover contacts which enabled me to arrange April meetings in Riyadh, Saudi Arabia with Al Saif Motors (Land Rover and Jaguar importers) and with Al Tayer Motors in Dubai (Land Rover importers). Also in April I arranged meetings with Galadari Automobiles (Ford), Al Ghandi Auto (GMC Iveco), Arabian Automobiles of the AW Rostamani Group (Nissan) and Trading Enterprises of the Al-Futtaim Group (Volvo).

Gordon Roscoe and I had been informed that the main way to establish a business in the UAE was to form a joint venture arrangement and to have a physical presence in either Dubai or Abu Dhabi. So during our visit we looked into a few possible options for renting office space in Dubai. There was an incredible amount of office building work going on in Dubai in the late 1990s. Finding an office would be easy.

During 1999 I became increasingly aware of serious financial concerns within JCG that were raised during JCG Directors' meetings. There was a high fixed cost in the business, mainly owing to the large number of consultants employed, and through the first half of 1999 there was an increasing need to retain existing business and generate more business. The situation didn't improve and by July 1999 I was told that to reduce costs I would have to cancel my Regus office contract as soon as possible. I was now unsure whether I had a future in JCG.

What turned out to be my last overseas business trip for JCG took place in September 1999. This would be an opportunity to obtain training consultancy business in the UAE. I arranged a number of meetings with vehicle importers in Kuwait which would be followed by a series of follow-up meetings with vehicle importers in Dubai. I travelled with Gordon Roscoe.

Our meetings over the next two days were with vehicle importers that had expressed interest in what we had to offer, especially the management training courses. We again met Chris Preston of Trading Enterprises and we were shown Al-Futtaim's excellent training facilities in Dubai that we could use. We again met Al Tayer Motors and Al Ghandi Autos and the next day another meeting with Brian Elsmore of Galadari Automobiles. We also met George Bizri, Vice President of Al Habtoor Motors, the Mitsubishi importer.

During these follow-up meetings in Dubai Gordon presented specific training proposals which were well received. What we had to do was demonstrate JCG's firm commitment to these automotive companies by setting up a physical presence in Dubai and with it at least two consultants that would be permanently based there to deliver the training courses.

Gordon and I had already discussed the possibility of relocating two of our consultants to the UAE for two to three years. We were informed on our earlier visit to Dubai that no other dedicated automotive consultancies were based in the UAE and that the UAE automotive companies wanted the best practice industry training programmes that British consultancies could offer. For JCG the opportunity was not only in Dubai but also in all seven Emirates, including Abu Dhabi, where these automotive importers also operated vehicle dealerships.

On Friday, 10th September at a Directors' meeting it was revealed the financial situation was serious and future prospects were not good. However, as requested Gordon Roscoe and I presented a proposal to establish a JCG office in Dubai from which to deliver training courses to a number of automotive importers and their dealer networks. In view of

JCG's financial position it was not surprising that Michael and Rob decided we could not proceed with the UAE venture.

As this was a fundamental decision not to progress a viable business opportunity outside the UK, and the main reason I joined JCG, I realized I had no future in JCG. On 8th November 1999 in a meeting with Michael we agreed I would leave JCG in January 2000.

Looking back I suppose one could say that JCG should not have aimed at developing its automotive consultancy business outside the UK, and that it should have kept its focus and attention on just the UK. However, Michael, Rob and I thought differently otherwise I wouldn't have joined JCG. Maybe with hindsight we should have gone straight to the UAE as an initial business target, because there was business ready to be had there. This business would have initially been the delivery of bespoke training courses, which were less liable to being hijacked. In addition, we could do it with English-speaking consultants we already employed.

Chapter 19

# Honda Motor Europe 2000–2009

**IN MAY 2000 I MET KEN KEIR,** Managing Director of Honda UK, at his office in Chiswick and we discussed some possible employment opportunities. Honda UK would be relocating shortly to new offices in Slough and at the same time Honda Motor Europe (HME) would be relocating to the same offices from Reading. Early in August I met Ken again to discuss a role in HME and two days later I met with him and Sally Gilliver, Head of HR, to discuss and agree my appointment in HME Network Development with effect from 1st September 2000. My role would be to help develop the HME car business in Europe. It would enable me to use my international experience and present me with another career challenge.

The 200-mile daily return journey from home in Dorridge to Slough was not a realistic option. If I wasn't travelling to Slough and back home each day I would be going to London Heathrow Airport, a 100-mile journey from home, for flights into Europe. So in agreement with Honda, which agreed to pay for my relocation costs to complete a home move nearer to Slough, we put our house in Dorridge on the market. Our next home would be located further south, just off the M40 motorway and no more than an hour's drive from the HME offices.

In my first days in Honda I learned I would lead a very small team ensuring HME and all companies within HME complied with all existing and future European Commission legislation relating to the automotive industry in the European Union (EU). So in my first week I travelled to Brussels with David King, HME Company Secretary, and Tadaki Kato,

HME Head of Legal, to meet Andrzej Kmiecik and Fergal O'Regan of HME's European lawyers, Van Bael & Bellis.

The purpose of the meeting was to discuss one important element of European competition legislation under review and for which HME's position would have to be developed. This was Block Exemption Regulations or BER for short. This trip to Brussels and meetings there would become a frequent occurrence for me as Van Bael & Bellis, JAMA (Japanese Automobile Manufacturers Association) and the European Commission were all located there. I will say more about BER and my involvement and meetings with these three organizations later.

At the end of September 2000 I attended the Paris Motor Show. This and other major European Motor Shows held in Frankfurt and Geneva were opportunities for me to have informal meetings with the senior management of the Honda National Sales Companies (known as 'Genpos' by the Japanese management) on relevant business matters.

I also knew there was another very common Japanese word I would use regularly in my communications with Honda's Japanese management. The word is 'san', which means all of 'Mr', 'Mrs' and 'Ms'. So Mr Minoru Harada, the President of HME, was Harada-san and Mr Tadaki Kato, HME Head of Legal, was Kato-san.

On joining HME it was agreed I should drive a few Honda cars that I wouldn't normally drive, simply to improve my Honda product knowledge. Early in October I borrowed a Honda Insight, one of the first petrol/electric hybrid vehicles available on the market at that time. This was a vehicle which I found with careful driving was capable of achieving just over 100 miles on one gallon of fuel. However, with its skinny wheels and relatively light weight I also found that the Insight was not the best for coping with some strong side winds that I experienced a few times on the M40 motorway.

The next car I borrowed and drove was at the other end of the model range, a Honda NSX super sportscar, with fantastic performance, and over the weekend of 28th and 29th October Annette and I did some house-hunting in Oxfordshire in the NSX. It was a great experience to drive an NSX even though it was not the most practical car, but then it was not designed to be. The NSX was launched in 1990 with advanced aerodynamics and styling apparently inspired by an F-16 fighter jet cockpit and with input from the late Formula One World Champion Ayrton Senna during the car's final development stages. This NSX became the world's first mass-produced sportscar to feature an all-aluminium body and was powered by an all-aluminium 3.0-litre V6 engine.

One of the houses we visited was in the small village of Ardley just off the M40, north of Oxford. This location enabled me to get to my office in Slough in about an hour, just using the M40 and a few short stretches of the M25 and M4. On 3rd November we accepted an offer on our house and within days made an offer on the property in Ardley, which was accepted. However, it was not until 25th January 2001 that we could move into our new home.

I referred previously to meetings with Van Bael & Bellis, HME's lawyers, to discuss European Block Exemption Regulations. What is Block Exemption? Here is a summary.

When the Treaty of Rome was signed by the original six members of the European Common Market in 1957, it instituted a free competitive market in those European countries. The treaty laid down a basic rule, Article 81(1) banning agreements which could have anti-competitive effects. However, Article 81(3) gave the EU Commission the power to exempt certain agreements, on condition that they respected certain requirements.

Motor vehicle distribution and servicing agreements were included as an 'Exemption en bloc' (from the French language) under the treaty in 1985 under Regulation 123/85. As a result motor manufacturers were allowed to create 'selective and exclusive' franchise networks. Regulation 123/85 allowed motor manufacturers to 'select' their dealers on the basis of both the number and quality (franchise standards) of dealer. Dealers in return were given 'exclusive' areas called territories, in which no other dealer of that franchise could be appointed.

Regulation 123/85 defined the way new vehicles are supplied in the EU countries. The rules are the same for all EU member countries. In 2000 there were ten member states. The BER was renewed by the European Commission in 1995 under Regulation 1475/95 with some minor changes for a further seven years.

With Regulation 1475/95 due to expire at the end of September 2002, in 2000 the European Commission had begun the process towards introducing a new BER regime by carrying out a major evaluation of the existing Regulation and by the end of 2000 produced and made available a detailed report. All interested parties including motor manufacturers were requested to submit a formal response to the European Commission's report before the end of January 2001, which would be followed by a public hearing in Brussels on 13th and 14th February 2001 to which all interested parties were invited to attend.

There would be far too much detail to describe here all the considerable amount of work carried out within HME and with Van Bael & Bellis to produce a formal HME position and submission. We also had meetings with JAMA so that this organization was aware of HME's views and likely position. On 15th January 2001 the following communication was sent to the European Commission. It contained a detailed eight-page submission.

> *Re Report on the evaluation of Regulation 1475/95*
>
> *Please find enclosed herewith the submission of Honda Motor Europe Ltd in response to the Commission's Evaluation Report on Regulation 1475/95. Honda would also like to inform you that it will attend the Commission's public hearing of 13 and 14 February 2001 as an observer. The persons attending the meeting will be John Sparrow, Tadaki Kato and Fergal Ó Regan.*

At the two-day public hearing in February 2001 the European Commission presented its Evaluation Report. The public hearing was attended by companies from all EU countries and all sectors of the automotive industry including motor vehicle manufacturers, parts manufacturers, dealer groups, suppliers, service providers, motoring organizations and associations. The hearing gave many of these interested parties in the wider automotive industry the opportunity to present their proposals for changes to be included in the successor regime to BER 1475/95, and responses to the Evaluation Report.

I recall a presentation made by Virgin Cars Ltd with a proposal that companies such as theirs should be able to sell new cars solely on the internet. At the time Virgin Cars Ltd was a new internet car retailer, established by Sir Richard Branson in May 2000, and part of the Virgin group of companies. During a break in the public hearing I had a conversation with Sir Richard, who explained he would like to sell new Honda cars on the internet and that this new retail channel would promote competition. I said Honda would not be willing to grant its vehicle franchise on an internet-only basis and that its franchise holders would be expected to provide their customers with premises and staff to test-drive cars, handle vehicle part exchanges, handle warranty work and provide servicing and repair facilities.

Honda could not envisage a situation where a franchise holder had no investment whatsoever in the franchise as proposed by Virgin Cars in their internet-only arrangement, and strongly opposed such a possibility in subsequent discussions with the European Commission, a position which the Commission accepted. Virgin Cars predicted it would sell around 24,000 cars in the first year, but by October 2000 it was reported that it had only sold 2,000 cars. By 2003, it was reported that Virgin Cars had only managed to sell 12,000 cars in total. In May 2003 Virgin Cars opened a car showroom in Salford, Greater Manchester. However, the venture was not a success and in December 2005, the company ceased trading.

Towards the end of February 2001 I made my first visit to Honda Motor Europe (North), or HME(N) for short, in Offenbach near Frankfurt. HME(N) was responsible for a region including Germany, Austria, Belgium and the Netherlands and was headed by its President Hiroshi Kobayashi. Arranging my first visit and meeting with Kobayashi-san was not easy. He did not think a meeting was necessary and told me there was nothing he could learn from me as I knew nothing about Germany. Unfortunately he didn't ask me where I had worked before Honda and I had to tell him I did know a good deal about the motor industry in Germany from my time in Rover Group and BMW Group. In addition, I had to tell him my visit was not only to brief him and his management on EU BER but also to discuss and get his team's input on my plans to develop a pan-European franchise standards programme, to ensure a consistent approach within the single EU market. Kobayashi-san reluctantly agreed for me to visit HME(N) which I did on 20th February 2001. However, I ended up discussing these subjects with the senior German management of Elmar Paltian and Lars Zeiler. It became clear that HME(N) was able to operate very independently from HME Head Office and that I would have to work hard to get agreement to my pan-European network development plans.

Honda Motor Europe (South), or HME(S) for short, was located at Marne La Vallee, to the east of Paris. HME(S) was responsible for a region including France, Italy, Spain and Portugal. On 7th March 2001 I had my first meeting at HME(S) offices with Sakata-san, President, Hitoshi Moriguchi, Vice President of HME(S), together with their Managers Andrea Biava and Benoit Beaufour. I soon learned that unlike HME(N), where Germany was the dominant market and did not greatly involve other HME(N) countries in major policy or strategy, HME(S) based in France did involve the other constituent countries of Italy, Spain and Portugal in its policy, strategy and operational activities.

There was a third European Region, HME(UK), which covered the rest of Europe including the UK, Norway, Sweden, Finland, Denmark,

Iceland, Ireland, Switzerland, Greece, Israel and Eastern European countries. As well as wholly owned national sales companies this region included all Honda's independent distributors in Europe.

In March 2001 I visited Mayer's Car & Truck Ltd, Honda's distributor in Israel, based in Tel Aviv. My visit was arranged at short notice at the request of Joe Bahat, a director of Mayer's. They were making a major investment for Honda in new showroom facilities in Tel Aviv and Joe Bahat wanted my advice and recommendations so they could comply with Honda's latest showroom décor and corporate identity.

On 15th June 2001 JAMA represented by JAMA's Director as well as Honda, Nissan, Mazda and Toyota representatives presented the 'JAMA Position on BER' to Eric van Ginderachter, of DG Competition (DG is Directorate-General) within the European Commission at its offices in Brussels. To give an indication of our thinking at that time I will mention the 'six essential features' that JAMA proposed should be included in any new BER regime.

→ *Maintain qualitative selection for authorized sales and after-sales services.*
→ *Maintain quantitative selection of authorized sales and service providers.*
→ *Allow linkage of sales and after-sales services, which is especially vital for small-volume suppliers.*
→ *Permit active sales outside areas of primary responsibility through personalized advertising.*
→ *Keep the current permissive rule on multi-brand sales, maintaining separation on all aspects that are visible to consumers, and support the value of the brand.*
→ *Maintain the prohibition on sales to resellers as well as the rules requiring sales to mandated intermediaries.*

These features will not mean much to people without knowledge of how the motor industry works. So I will just refer to 'qualitative selection', the first of the six features. This simply meant motor vehicle manufacturers should select their franchised dealers on the basis the dealer meets the manufacturers' defined franchise standards.

On 2nd July 2001 I made a presentation on the BER situation and JAMA's position to Minoru Harada, HME President, and his Senior Directors. The boardroom was full of all the presidents of the European Genpos and I think the only senior Europeans present were Ken Keir, Honda UK Managing Director; Chris Rogers, HME Head of Corporate Affairs; David King, HME Company Secretary; and me. After my presentation I asked if there were any questions. There was total silence and rather than saying thank you and finishing I asked again.

I now realized that unless Harada-san said something no one else present was going to say anything. Fortunately Harada-san who was sitting at the front of the room closest to me asked me to go back to my first slide and we proceeded to have a short question-and-answer session on the Block Exemption topic between ourselves, while the rest of the audience listened silently. Harada-san was also on the main Honda Motor Co board and so with his very senior position it was clear that in the presence of the HME President no other Japanese HME President would speak unless invited to do so.

For me Harada-san was a very approachable HME President and on some occasions I would go to his office and ask if he was available for a few minutes so I could update him on certain EU matters that affected Honda. I think he knew I was not there just to give good news but to tell him things honestly and exactly as they were. On one occasion I went to tell him that one of his direct reporting European Presidents was somewhat opposed to implementing one of our pan-European programmes and would he tell him to do it – as that person did not report to me. The response I got was that it was up to me to discuss and convince the President of that country to do it and thereby gain his agreement.

What I realized was that each European President of the Honda Genpos had a great deal of freedom to run their business very much as they saw fit, and so trying to develop and introduce pan-European programmes for HME became for me very time-consuming and often frustrating.

The Honda business culture was very different to what I had experienced in the BMW, where BMW's strategic direction and policy from senior management was a much more frequent part of normal day-to-day business, and not to be opposed or questioned. I think it would have been inconceivable in BMW at that time for a BMW Executive in a European country to oppose a pan-European programme directive from BMW Head Office in Munich. And yet I was here at HME's European Head Office in Slough having to convince the Honda President of a country of the need to agree to implement a pan-European Honda car programme, and one that HME was required to implement consistently in all EU countries, because we were operating in the single European market. In Honda everyone had great freedom to do their job to the best of their ability without much restriction or dictation from top management.

Within HME I learned that only Honda UK had introduced a franchise standards programme and measured to ensure compliance. On mainland Europe there were various lower-level standards programmes in existence, with variable content and consistency. As part of the approach and position now decided on BER within HME and JAMA it was agreed within HME that I would develop a comprehensive and consistent set of franchise standards to be adopted by all Genpos and distributors within the EU for their Honda car dealer networks.

At the same time I would also coordinate the development and introduction of a franchise standards programme for Honda's independent distributors in the EU which would also require compliance with the BER. As a result any new distributor that Honda wished to appoint would be required to meet a full set of defined and documented standards including, sales, service and parts facilities and operations, finance, staffing levels and training. These standards would also include a requirement for Honda car dealers to implement HME's corporate identity programme.

In 2001 Honda UK was considering updating its dealer corporate identity and so to ensure there would be a consistent approach for all Honda markets in the EU, I started to look at this subject. I knew a high-quality corporate identity programme would become an important and very visible franchise standard and a standard that would aim to identify and promote Honda as a premium car brand across Europe.

In July 2001 I met Eric Tienvrot, a director of Arlux, one of two suppliers to Honda UK, on its corporate identity (CI) programme. I already knew Arlux, having visited its Head Office and manufacturing plant in Nantes, France in 1982. I wanted to explore the best way to implement a pan-European CI programme and to consider whether Arlux had the capability to be a potential supplier for HME in Europe, as well as the UK.

During August 2001 I had a number of meetings with Honda UK management on the subject of franchise standards since the UK's programme was likely to form the basis for a minimum franchise standards programme for all HME's markets in the EU, and for those markets outside the EU. In 2001 there were fifteen countries that comprised the EU. These were Belgium, France, Germany, Italy, Luxembourg, the Netherlands, Denmark, Ireland, UK, Greece, Spain, Portugal, Austria, Finland and Sweden.

However, there were many other European countries scheduled or aiming to join the EU so my pan-European programmes had to take these markets into account. In May 2004 a further ten countries joined the EU. These were the Czech Republic, Hungary, Poland, Slovakia, Slovenia, Estonia, Latvia, Lithuania, Cyprus and Malta. In January 2007 Bulgaria and Romania joined the EU to bring the total EU membership to twenty-seven countries.

To understand the actual Honda dealer standards situation in some of these aspiring EU countries I visited Hungary at the end of August

2001. After an initial business meeting in Budapest with Nozawa-san, President of Honda Hungary, and Janos Reisz, General Manager, I visited a number of Honda car dealers in Békéscsaba, Szeged, Kecskemét, Jászberény and Szolnok with Tamas Datscsinsky, Dealer Development Manager. These dealers were in key locations in Hungary and although small in size by UK standards I judged that these dealers were capable of meeting the sort of minimum franchise standards I was developing for HME's markets. This visit to Hungary was a good example of Honda's business philosophy of 'go to the place' to understand and then make decisions.

On 11th September 2001 now known internationally as '9/11' four coordinated attacks were made on the United States. Two American Airlines planes were crashed into the North and South towers of the World Trade Center complex in New York City. Honda immediately responded by banning all staff travel by air. As a result on 18th September Kato-san, Head of HME Legal, and I travelled by Eurostar from London to Brussels for a number of Block Exemption meetings with Van Bael & Bellis and JAMA. Our journey was not without incident. Shortly after entering the Channel Tunnel the train we were on stopped. We didn't move for forty minutes and there was no communication to tell us why we weren't moving. Also we were unable to use our mobile phones to find out what was going on.

This unexpected long stop caused a good deal of concern amongst the passengers some of whom thought the stop was terrorist-related. When we finally arrived in Brussels we were told the train in front of ours had broken down. I had to make two more journeys to Brussels by Eurostar before Honda management allowed staff air travel once again.

In December 2001 I attended a first cross-functional Sales, Engineering and Development (SED) meeting at Honda's manufacturing plant in Swindon. This was set up to ensure a good exchange of information between HME Sales, HME Manufacturing and HRE (Honda Research and Development Europe) and to discuss areas of common interest at senior management level.

I was the main representative for HME Sales, Mike Godfrey for Manufacturing and Adrian Kilham for Research & Development. I would brief the others on the prevailing sales situation in Europe and on most occasions Steve Oliver attended to outline the Honda UK sales and business position.

The official Honda definition of SED is as follows.

*A perfect balance that leads to products with originality and vision. SED refers to the three corporate functions that are vital to Honda's core activities: Sales, Engineering, and Development. This system forms the basis for the product development process, capitalizing on the specific strengths of specialists in each of these respective spheres, while maintaining close integration among all three groups. This unique combination of individual achievement and mutual cooperation is the unified driving force that propels ideas from development through to production to create products that will please both customers and society as a whole.*

In December 2001 I also attended the first Project ET meeting with Ken Keir, David King and Chris Rogers. ET was short for European Transformation, and the project was established to determine a strategy and action plan for growing Honda's European car sales. By 21st January 2002 we had put together a detailed presentation setting out Honda's strengths, weaknesses, opportunities and threats in Europe. In addition, the presentation included HME's strategic direction, HME's pan-European structure, roles and responsibilities, and NSC structure, roles and responsibilities. The presentation summarized the measurement criteria for achieving sales, marketing, supply and after-sales objectives.

Under the heading 'Structure' we decided that 'Europe needs 1000 quality car dealers doing 250,000+ and capable of expansion'. HME's ambition was to grow to achieve 500,000 car sales per annum in Europe. Unfortunately in the following years HME never came close to realizing those sales volume ambitions. In 2017 Honda's total car sales for the full year were 140,343 representing just 0.9% of an increased 15.63 million car registrations in the total EU and EFTA countries. There are many reasons for these low sales volumes. These include an extremely competitive European car market, a Honda model range reliant on Jazz, Civic and CR-V, a reliance on the retail customer and increasingly competitive and attractive offerings from other motor manufacturers, such as Hyundai and Kia.

The first few months of 2002 were taken up with more 'Project ET' and BER meetings. On the latter we were now entering a period where all Honda's European Genpos had to be involved, primarily to ensure they were able to brief their dealer networks on the various areas where they would have to comply. These areas included the implementation of minimum franchise standards and various sales, after-sales and business obligations. To make sure all areas of HME were involved and able to take appropriate business actions I established a BER Project Team in HME. In addition to me as project leader the HME team comprised David King (Company Secretary), Richard Knill (Business Administration), T. Kato (Legal), T. Kono (Sales), Albert Erlacher (Pan-E Marketing), Dick Klein (Service), Chris Rogers (PR) and Ian Crawford (Parts). These persons made up the main working group. In addition, HUK was represented by John Kennedy (Network Development) and Steve Oliver (Business Support). HME(N) was represented by Elmar Paltian, HME(S) by Bruno Haas and Pascal Delavenne, and the Nordic region by Jonas Lindow.

At the end of February and in early March, along with Chris Rogers, Dick Klein and Kato-san, I attended meetings at the offices of HME(N), HME(S) and the Nordic region to ensure each local team was briefed on the specific actions they would need to take and following all this to produce a detailed deployment plan. At the HME Presidents' meeting in April 2002 I presented a status report on the BER situation to make sure all the HME Presidents fully understood the business requirements and obligations with which HME would have to comply in a consistent way in the single EU market, i.e., all EU markets. I could not assume they would all be briefed in a consistent way locally.

In March Annette and I had our first holiday of many in Dubai. I refer to this mainly because I had been invited to play golf at the Abu Dhabi National Golf Course by Steve Kiggins, who was working in the oil industry there. Steve's wife Sheila was a relative of Paul and Judy Smith, our friends and our ex-neighbours in Dorridge, who were also on holiday in Dubai. The Abu Dhabi National is a magnificent golf course with a large metallic falcon spanning the façade of the clubhouse, reflecting the traditional pastime of falconry. Its outstretched wings form the roof of the building. I remember the superb condition of the fairways. Each fairway was like that of the surface of the putting greens – it seemed a shame to take a divot when playing a shot.

Steve Kiggins arranged the game and with his colleague Keith Torrance drove Paul Smith, me and our wives from Dubai where we were staying to Abu Dhabi, a journey of around 85 miles. We were enjoying the game, however, on the 10th hole I thought for a few moments I might die there. I had played a shot into the green but the ball rolled through the back of the green into some palm trees. As I went to play the next shot the sharp pointed tip of one of the fronds (leaves) pierced the back of my left wrist and blood immediately began spouting out. Fortunately Steve told me to lift my hand up above my head while pressing a finger into the wound. After a few minutes the blood stopped and a few minutes later after the shock had subsided we continued to finish the round. My polo shirt was covered in blood and as we met our wives at the clubhouse they wondered what had happened to me. I think my golfing friends also had some of my blood on them. Since then I have been very careful on the occasions when I have been near palm trees.

On 8th May 2002 Senior Management from Honda, Nissan and Toyota held a meeting in Brussels with Mario Monti, the European Commissioner for Competition. This meeting had been arranged to enable these three Japanese motor manufacturers to put forward their views on the nature and content of the new BER. Harada-san, HME President, together with Chis Rogers, Kato-san and I represented Honda and a short presentation was made by Harada-san and by senior directors from Nissan and Toyota.

However, the main response from Mario Monti was that the motor manufacturers were very large international companies which were perfectly able to meet and implement the European Commission's regulations aimed at increasing the level of competition in the EU for the benefit of consumers. He didn't seem to take in the fact presented to him that the motor manufacturers would automatically aim to be as competitive as possible, without the need to comply with more bureaucratic legislation.

All through 2002 many meetings were held with HUK, HME(N), HME(S) and the Nordic region to agree a set of minimum franchise standards for the HME's car dealer networks. Eventually on 31st October 2002 I was able to present to Harada-san, HME President, the proposed HME minimum franchise standards programme, together with details and explanatory notes of the standards assessment procedure to be followed by Genpos and distributor field staff carrying out the dealer standards audits. The programme was finally signed off and agreed by Harada-san, and Saito-san, HME Vice President, at a meeting on 7th January 2003.

I had a first meeting with Marcus Imfeld of Westiform in August 2002 to discuss the development and implementation of a consistent corporate identity signage programme for HME's car dealer networks. I also needed to get a clear idea of the cost of making and installing the signage, and to ensure I would have one consistent set of materials and costs from the two suppliers.

Dealer corporate identity would be an important minimum franchise standard that I wanted to introduce across Europe. Both Arlux and Westiform wished to be suppliers to HME in Europe and I judged that together they had the capability of developing and implementing a consistent and high-quality signage programme. With Arlux based in Nantes France and Westiform based in Ortenburg in southern Germany I calculated it would be possible to give each company roughly half of HME's car dealer networks. This would give both companies a similar financial opportunity and business potential, which I also thought was important.

My proposal was that Arlux would cover HUK, HME(S) and the Nordic region, while Westiform would cover HME(N) and Eastern European countries. On 3rd October 2002 I made a detailed corporate identity presentation to Harada-san and his HME Vice President Saito-san to set out my proposals for a new high-quality corporate identity programme for the European car dealer network, and one which would raise the image of the Honda car brand and the Honda car dealers in a consistent way across Europe.

I explained in the presentation that in order to ensure the European dealer network would be willing to make the required investment, that HME Genpos and distributors should contribute 50% of the cost of the programme and the dealers the other 50%. In addition, I said that I would need to recruit a dedicated Corporate Identity Manager to manage and oversee the development and implementation of the project and to be the main point of contact in HME for all the Genpos and distributors on this subject.

At the time there were around 1,500 Honda car dealers in the HME region. I was delighted to get Harada-san's approval of the project, of the estimated total costs and of the proposed apportionment of the costs. However, he asked why I needed to recruit someone to manage the programme.

Only after I explained the necessity and importance of the role did Harada-san give his agreement to the recruitment of a Corporate Identity Manager, a role that would report to me.

I wasted no time in arranging the recruitment process with the HR department. This involved advertising the role internally and then if necessary externally. At the same time I visited Arlux and Westiform, to inform them of HME's plans to introduce a new corporate identity programme for its European car dealers and of the roles and responsibilities of these two suppliers in working together to develop and implement the programme over a two- to three-year period. The response from both companies was extremely positive.

I completed the interviewing process and after careful consideration and agreement with HR I phoned Doris Pfingstner to tell her we were going to offer her the Corporate Identity Manager's job. After I received her verbal acceptance we sent her a formal offer letter. Doris was an ideal person for the job. I was looking for someone with a strong personality, positive attitude, clear understanding of the role, a great communicator and someone to whom I could confidently delegate all key aspects of the job. And not least, someone who I was sure could get the programme developed and implemented on time and within the budget. I was very lucky to find Doris, who I was sure met all these requirements. In addition, I wanted someone who spoke German or French fluently as this would help in communication and negotiation with the main European markets. Doris was an Austrian national fluent in English and German, and I knew that a fluent German speaker would be extremely beneficial in communication and in achieving a good working relationship with both top and operational management in HME(N). With key issues of the franchise standards and corporate identity programmes now decided, 2002 ended for me on a very positive note.

In January 2003 I emailed the proposed minimum franchise standards programme to all European Genpos. That month I also held the first joint meeting with Arlux and Westiform management at the HME offices in Slough to discuss costs and pricing of the CI programme. I confirmed Doris Pfingstner would manage the programme and on 3rd February 2003 Doris started work in HME.

One other business topic that was occupying more of my time was 'corporate sales'. Within the Honda Motor Europe region at that time only Honda UK regarded corporate sales or fleet sales as an important activity within its total car sales business. In mainland Europe the Genpos' business was almost totally based on vehicle sales in the retail market, or sales to individual customers. During 2001 and 2002 I was approached by a number of leasing companies, some of which I had worked with earlier in my Land Rover and BMW days. These companies were becoming more interested in Honda's vehicles, often because of their experience of buying Honda cars in the UK.

So in agreement with HME top management I took responsibility on a pan-European basis for corporate sales activity in HME. I discussed this subject with Harada-san, who told me that selling vehicles to the corporate market involved giving big discounts and that HME would be better off selling its cars 'one by one' to individual customers. I remember saying that as HME had an ambition to grow and sell 500,000 vehicles a year in Europe it could not ignore the corporate car sector, as this accounted for at least 30% of all new cars sold each year. I also said I thought it realistic to achieve an incremental 10,000 new car sales in the first few years at no worse than break-even, sales of new cars which would be very useful in contributing to the meeting of fixed costs of Honda Manufacturing in Swindon.

Corporate Sales now became another important business agenda item in my meetings with Genpos and distributors, but establishing a pan-European corporate sales procedure in HME became one of the most difficult things to achieve, not least because of the difficulty in getting top management in all the Genpos to accept the need to establish a human resource to exploit this market sector and to set realistic discount terms.

In the week of 19th May 2003 I visited the Honda Genpo management in Norway, Denmark, Sweden, Estonia and Finland. On the Monday I flew to Oslo and visited seven Honda dealers with Kare Filseth, the General Manager of Honda Norway. The next morning I flew to Copenhagen and visited three Honda dealers in the city area with Mikael Larsen, the General Manager of Honda Denmark. That evening I took the forty-minute train journey from Copenhagen to Malmö across the Oresund Bridge and dined with Ola Davidsson, Vice President of the Nordic Region, and Jonas Lindow, the General Manager of Honda Sweden. The next day I visited some dealers in the Malmö area with Jonas and flew later to Tallinn, the capital of Estonia, where I met Tonu Vahtel, the General Manager of Honda Estonia. On the Thursday Tonu and I visited dealers in Tartu, Viljandi, Parnu and Tallinn.

At 6:30 p.m. I took one of my more memorable flights from Tallinn to Helsinki. This was a twenty-minute flight by helicopter which was memorable not only because the helicopter seemed to me to be flying very close to the water the whole way, but also because I thought it was an unbelievably noisy journey, even for a helicopter, despite the fact the few passengers and I were wearing ear defenders to reduce the noise. On reaching Helsinki I dined with Jyrki Makinen, the General Manager of Honda Finland, and the following day visited Veho in Espoo and Helsinki. That evening I flew back to London.

This was a typical very full and busy week's work but a very rewarding one that enabled me to get to know the HME Nordic management, for them to get to know me and my business objectives, to discuss areas of common interest face-to-face at the same time, and to better understand the standard of the dealer representation in these five countries. Each time I visited a country I asked the local management to include some dealer visits in the schedule, not only to see for myself the local standard of dealer representation, but also to communicate to the dealers as a senior manager from HME Head Office our plans and intentions on subjects such as business objectives, sales plans and minimum franchise standards.

I found it was not often that anyone from HME Head Office included Honda car dealers in their schedule when visiting the Genpos' offices. However, I made a point of including some Honda dealer visits on my business trips, especially if I hadn't been to that country before.

The development and implementation of a new corporate identity signage programme for HME's car dealers for the whole of Europe also raised questions within HME about the future corporate identity signage for Honda's motorcycle dealers and power equipment dealers.

The Headquarters of Honda's motorcycle business in Europe, Honda Europe Motorcycle SRL, or HEM, was in Rome, while Honda's power equipment business in Europe, Honda Europe Power Equipment SA, or HEPE, was in Ormes, France. There were relatively few situations in Europe where a Honda car dealer also had Honda motorcycles and/or power equipment products in the same showrooms or premises. However, with the introduction of a new CI for the car dealers there was a need to ensure there was an approved company CI signage where a dealer represented Honda cars and Honda motorcycles and/or Honda power equipment products.

In order to find a satisfactory CI solution to these multiproduct situations in May 2003 I met Christophe Baillien and Tristan Durivault of HEM in Rome. As a result of the meeting the term 'hybrid dealer' was agreed to describe these Honda multiproduct dealers and it was further agreed that appropriate CI signage would be developed within the HME CI programme for these 'hybrid dealers', all of which would follow the red and silver colours of the HME car dealer programme.

An interesting event took place during my visit to Honda Automobiles Suisse SA (HASSA) on 18th June 2003. My visit was to discuss the new corporate identity programme as well as other operational programmes with this independent distributor. However, during the afternoon meeting the local management told me that at 4 p.m. that afternoon Honda had acquired HASSA, and that it was now a Honda Genpo.

The subject of CI up to now was limited to the new exterior signage that would be installed on the showroom fascia of the dealership. However, questions were being asked both in HME and in the Genpos about the minimum standards of décor that HME would require dealers to adopt inside their premises. Up to now there was no consistent interior CI or décor for HME car dealers. So this subject was now included into an enlarged CI programme and the first HME interior décor meeting took place at HME's offices on 8th July 2003. This meant there was an extra piece of responsibility and more work for Doris.

On 24th July a second interior décor meeting was held at HME offices, at which it was decided there would be a number of mandatory décor items that would become the minimum standard, but there would also be some optional items of décor that the dealer could adopt. It was also agreed that a consultancy, experienced in developing and implementing a branded car dealer interior décor programme, was necessary and that we should sound out potential companies and invite them to present their case for being selected.

The one consultancy that I knew well was Design Forum based in Dayton, Ohio, USA, a firm that had successfully developed the overall Land Rover Center concept for Land Rover North America in the 1990s, including interior and exterior décor, and with it a unique brand experience for the customer. Design Forum also developed and implemented Honda's corporate identity programme for its North American car dealer network. Interestingly that programme was based on the colour blue, whereas the HME colour scheme for its dealers in Europe was red and silver.

On 31st July I phoned Lee Carpenter, Design Forum's CEO, and explained to him what we were looking to do and to see if he would be interested in pitching for the interior décor project. If so he would need to make a presentation to HME management, probably in early September. I was delighted to hear his positive response and that Design Forum would welcome the opportunity to discuss the matter before presenting its proposals. Accordingly a video conference was arranged and held on 15th August.

On 5th September 2003 a number of potential suppliers that were interested in the HME interior décor project, including Design Forum, made presentations to HME management, including Doris and me. Following a review of the candidates Doris and I agreed Design Forum would be our preferred supplier. On 16th September Doris and I made a presentation to Harada-san, Ken Keir, Uchida-san and Kobayashi-san outlining a proposal to appoint Design Forum to develop an interior décor programme for HME's dealer network to complement the new exterior corporate identity signage. In addition, we agreed with Design Forum that they would develop proposals for new Honda dealer premises that could be followed in situations where a new dealer facility was being built.

On 9th October 2003 I attended a meeting at the HME offices to discuss the start of another important project. This was named Project R81 and concerned Russia. The Russian market had been managed from Japan as part of the Asia region but now it would become part of an enlarged European region. However, a significant operational change would be the creation of a Honda Genpo that would manage and develop the car business from offices in Moscow.

The following day I convened a meeting with HME management to discuss the operational actions that HME would need to take with the distributors in those countries that would become new member states in the EU on 1st May 2004. These new member states would be the Czech Republic, Cyprus, Estonia, Latvia, Lithuania, Hungary, Malta, Poland, Slovakia and Slovenia. In practice Honda distributors in these markets would need to comply with all the EU legislation affecting the automotive industry in the single market, including Block Exemption Regulations (BER). We also needed to consider the fact that two more Eastern European countries, namely Bulgaria and Romania, were to join the EU on 1st January 2007. The latest part of EU enlargement saw Croatia join the EU as its twenty-eighth member state on 1st July 2013.

In November 2003 I made visits to Honda's National Companies in Prague in the Czech Republic and Budapest in Hungary, primarily to brief the local management on BER and to discuss and agree actions they would need to take to comply with these EU regulations, and

HME's operational requirements, including minimum franchise standards. These visits were immediately followed by a meeting in Malmö, Sweden to discuss what was named the HME Nordic/Baltic Project. In practice Honda's business in Estonia, Latvia and Lithuania would now be managed more closely by Tonu Vahtel in Estonia to ensure compliance with EU legislation, and like the Czech Republic and Hungary to meet HME's operational requirements.

On 18th November 2003 I visited Warsaw, the capital of Poland, for the first time to discuss BER and other business matters with Honda Poland's management. The following week I made another visit to Helsinki and Tallinn to discuss BER. Once again this involved me taking the noisy helicopter return flight to Helsinki before flying to Dusseldorf. The next day I had a full schedule of dealer visits with Lars Zeiler, which included Honda dealers in Dusseldorf, Cologne, Wuppertal, Dortmund, Bielefeld and Gütersloh.

In July 2003 Honda UK announced a competition for the HUK and HME staff, where those who wished to take part had to estimate the UK's total new car market for the month of September, the month with the new registration plate and the largest month of the year for new car registrations. In early October Ken Keir greeted me one morning by saying I was being transferred to HUK as Forecasting Manager. He followed these words by saying I had won the competition by submitting an estimate closest to the actual registration total. The prize was a long weekend in Europe. On Friday 28th November 2003 Annette and I flew to Vienna for the weekend break courtesy of Honda UK.

The main item in my diary for December 2003 was a four-day trip to Moscow with Ben Morgans, to progress our involvement on the Russian Project. Ben was a young manager who would now be the main HME contact on car sales in Russia. On 8th December we flew to Moscow and stayed at the World Trade Centre Hotel. The next morning we visited Honda car dealers Alan Z, FK Motors and Aoyama, as well as competitor car dealers representing Toyota/Lexus, Audi/VW and

Mercedes. The next day we visited Flight Auto (Honda) and dealers representing Skoda, Mitsubishi and Toyota.

It was clear also that the car dealers we visited had invested and developed premises on very large sites, so I assumed the availability and cost of land must have been very attractive for the dealer owners. Ben and I spent 11th December at Honda's Moscow office mainly discussing dealer network development with the local Honda Russia management.

The last major issue in December was the annual department budget meeting, held with Koike-san, who was responsible for business administration. This meeting was where forward budgets had to be justified and if necessary negotiated and agreed with him. In HME each budget year was called a 'Ki' and for 2004 the budget year was 81Ki. Why Ki? I can't remember. During my meeting Koike-san debated and questioned almost every budget figure for my department. In the end, and having prepared well, I was very pleased that my department's budget for 2004 was agreed intact and as submitted. With budgets agreed 2003 ended for me on a positive note.

At the end of January 2004, I arranged a meeting at the HME(S) office in France attended by HME(S), HME(N) management, Howard Thomas, Senior Vice President of LeasePlan, and myself. In my previous meetings with LeasePlan it had expressed interest in a pan-European vehicle supply arrangement with Honda. Such a supply arrangement already existed with Honda UK.

I arranged and attended many meetings over the next few years not only with LeasePlan but other major fleet users, as a few of Honda's vehicles, such as the CR-V, became more interesting to them. Unfortunately it took a long time for HME's European Genpos to be positive about the sales opportunities as Honda UK already was.

On 9th February 2004 I travelled with Ben Morgans to Minsk, the capital of Belarus, to meet with the directors of Paritet Service, the Honda distributor. Not surprisingly for winter in Belarus it was very cold and much snow had fallen when we arrived. In their presentation the next day it was clear this was a very small-volume market opportunity for Honda. However, our purpose was to discuss our intention to develop all the distributor markets as far as possible. The distributor's service facilities were very small, on a trading estate, and not as good as I was expecting but probably adequate for their existing business needs.

During the following day we visited dealers in Minsk that represented Nissan/Subaru, VW, Mazda, Peugeot, Toyota, Citroen, BMW and Skoda. It was clear that many of these competitor dealers provided a much higher standard of representation than Honda's. In particular I remember the VW dealer had underfloor heating in its service workshop, so in the harsh winter conditions in Minsk the technicians had excellent working conditions. I wasn't expecting to see that.

In February 2004 I visited Bernhard, the Honda distributor in Reykjavik, Iceland, for the first time. As with other visits to Honda's distributor markets, the purpose was to discuss the development of the Honda business, Honda's new standards for its distributors and dealers and for me to understand and learn about the car market and local competition. On this visit I flew with one of Bernhard's directors to Akureyri, the second-largest town, on the north coast to meet the directors of Holdur. This dealer, owing to the relatively small car market in the area, and as one of only a few dealers in the town, had been able to take on VW, Audi, Skoda, Mitsubishi and Peugeot franchises as well as Honda. All of the franchises were housed and operated in one showroom and one service workshop, and as far as I could see, the Honda car franchise was promoted and run in an acceptable way for this location.

Through February, March and April of 2004 we held a number of interior décor meetings with Design Forum and Genpo Dealer Development Managers to finalise the programme and these culminated with the identification of five dealers to pilot the programme's introduction. These dealers were Honda Zurich in Switzerland, Ibertechno in Barcelona, Spain, Japauto in Paris, France, Honda Padua in Italy and Boras in Sweden. In May Takagi-san agreed the plans to introduce a new Interior CI programme in all HME markets and I communicated this programme by email to all Genpo Presidents.

Early in April Doris confirmed she would be going back to Austria with her partner Martin and would leave HME at the end of June 2005. Doris had made a very significant contribution to the development, introduction and management of the new CI programme in Europe, and fortunately there was enough overall momentum on the programme for her successor to inherit. In July 2005 I interviewed Emma Hawkes, who was working in HME(N) in Germany but who was looking to return to the UK. Emma had the necessary skills, personal qualities and ability. In addition, as a German speaker she had the benefit of having good working relationships in HME(N), and so being the best candidate I offered her the job which I was very pleased she accepted.

In August 2005 I visited Tesco's Head Office in Cheshunt, Hertfordshire, to meet John Browett, Director of Tesco UK and prior to that Chief Executive of Tesco.com. I had met John at an ICDP conference earlier that year where he was a guest speaker. I sat next to John at that conference's lunch and he said he would like me to arrange for Tesco Main Board Directors to visit Honda's manufacturing plant at Swindon, so they could learn about Honda's 'Just in Time' process. I said I would be pleased to arrange a visit, but in return I would like to visit one of his supermarkets to see what I could learn about Tesco's retailing activity. We both agreed we might learn something from each other's business.

The visit I made to Tesco in Cheshunt with Steve Oliver, Head of the Customer in Honda UK, was very enlightening. Most of our visit was spent on a tour of Tesco's supermarket, with John Browett explaining in great detail aspects of their business. 'Just in Time' was important because there was relatively limited warehouse space and perishable food such as bread and milk went almost straight from outside into the supermarket for sale.

I remember John saying revenue per square metre was an important measure for them and that if one product was selling slowly its shelf space could be reduced and a faster-selling product would be given more shelf space. I also remember him saying the total investment in a Tesco supermarket was similar to a Mercedes-Benz dealership. The main difference was that a Tesco supermarket had much more land given over to car parking and that the building was much more functional. It would not have expensive flooring and ceilings that the Mercedes car dealer was required to have. Customers did not look up. The Mercedes dealer would have more expensive premises per square metre than the supermarket, but invariably not enough customer car parking. The revenue-per-square-metre business measure has stayed with me ever since.

John also told me they had thousands of customers visiting each supermarket every week and because they have a Tesco 'Clubcard' they know a great deal about their customers. They know what they buy, when the buy, how much they spend. However, he said even though the car dealer may not even have a hundred people visit the dealership each week, it will know much less than the Tesco supermarket about each person making the visit.

I heard from John on our visit that the Cheshunt supermarket was selling almost every type of product, including food, clothing, all types of white goods, televisions, mobile phones and so on. However, when I asked him about selling new cars he said that would be unlikely as Tesco could not control the motor manufacturer. John's comment I think meant Tesco could 'control' all other suppliers, presumably in terms of product quality, specification and above all price.

Interestingly Tesco did set up an online used-car operation, 'Tesco Cars', in April 2011 but only one year later it closed with the following statement. 'Following a review of the business model we and Carsite, our partner, have decided that we cannot offer customers a satisfactory range of vehicles and as a result, have decided it is right to close the business.'

It took much longer to arrange the Tesco Main Board Director visit to Honda's manufacturing plant at HUM Swindon. However, on 24th March 2006 a group of directors including David Potts (in 2022 CEO of Morrisons) were given a tour of the car manufacturing plant. I was told they all found the visit to HUM very beneficial.

In September 2005 I visited Honda Motor Russia (HMR) in Moscow with Dale Butcher, Director of Inchcape plc, to discuss and if possible progress the inclusion of this major international motor group into the Honda dealer network in Russia. Earlier in the year Ken Keir and I met with Peter Johnson, Inchcape's Chief Executive, and we agreed we should have Inchcape representing Honda somewhere in Europe, as at that time that had still not happened. It had been left to Dale Butcher and me to find a significant opportunity, and Russia was one jointly agreed possibility. At our meeting Kato-san, President of Honda Motor Rus, agreed in principle to include Inchcape plc in the Moscow dealer network, subject to Inchcape identifying a suitable opportunity in a required location and then proposing an investment/timeline that would meet with Kato-san's approval.

A further meeting took place in Moscow with HMR management in November 2005 to progress discussions, when I was joined by Martin Taylor and Kieran Godwin of Inchcape. In early December a meeting was held with Andre Lacroix, the new Chief Executive of Inchcape, Dale Butcher, Ken Keir and me to review progress. At that time all seemed to be going well towards Inchcape representing Honda cars in Moscow. During the continuing discussions between Honda and Inchcape it was a big surprise and disappointment when I learned from Dale Butcher that the President of Honda Russia had rejected Inchcape's proposals for a Honda dealership in Moscow. The President of Honda Russia had the final say on dealer network decisions so there was nothing I could do.

In December 2005 the CI programme took on an additional dimension when HME top management agreed that the programme should be expanded to be a Honda brand development project, including dealer premises layout as well as exterior and interior corporate identity. This decision was based on the fact that a number of new exclusive

Honda car dealerships were being planned around Europe and the dealer management were requesting and expecting full details of Honda's dealer facility and premises requirements. However there was no HME document available on these subjects at that time.

The first meeting was held in December 2005 at HME offices in Slough with Design Forum's senior management and Hakuhodo management. Hakuhodo, a global marketing consultancy, was Honda Motor's brand consultancy. We were now expected in HME to bow to the views of Honda Motor's consultants Hakuhodo, and to accept what Hakuhodo managers had to say on premises development and facility layout. It increasingly concerned me that there were now varying opinions, including those of the two consultancies and my own. I was no longer in control of this project.

Not long after I was told to inform Design Forum to stop their work on the project. This was a very disappointing way for me to end my association with Lee Carpenter and his team as I broadly agreed with their design proposals. However, the decision for them to stop work was not a surprise to me because Hakuhodo clearly didn't want Design Forum to be involved and had clearly gained Honda top management approval to this. Over the next few months the dealer design project was gradually taken over by Komuro-san, HME's Service Director.

As a result I spent much of the rest of 2006 progressing the introduction and measurement of HME's updated franchise standards programme and working on HME's position on the new EU BER that would come into being for the motor industry in May 2010. The latter subject started to involve more frequent visits to Brussels to meet with HME's lawyers Van Bael & Bellis and with the JAMA working group that included management from Toyota, Nissan, Mazda as well as JAMA management. My main contact at Van Bael & Bellis was Andrzej Kmiecik now that Fergal O'Regan had left during 2006 to become Head of Unit at the European Ombudsman. Meetings were also held with Paolo Cesarini of DG Competition and his team at the European Commission.

What occupied me in 2006 continued into 2007. There were more internal meetings and visits to Honda Genpos to discuss franchise standards and the implications on them of revised EU BER. Meetings continued with Van Bael & Bellis, the JAMA working group and with Paolo Cesarini and his team in the European Commission. On 29th June 2007 I sent to DG Competition in the European Commission Honda Motor Europe's response to the very detailed questionnaire we had to complete. This and the responses of all other motor vehicle manufacturers would then become subject to review by the European Commission and lead to more discussions in 2008.

On 21st November 2008 I attended a briefing by Takagi-san, HME's President. What he had to say resulted mainly from the financial crisis that hit the world's stock markets. Many international businesses, including motor vehicle manufacturers, bore the scars from the financial crash for the next four years. In that meeting Takagi-san announced that car production at HUM Swindon would stop during February and March 2009. In addition, Honda would cease to operate its Formula One racing team and other measures were being considered to protect Honda Motor Co's global business. The decision to close its F1 racing team seemed strange to me. Yes, 2008 had not been a good year but some of us that took an interest in Honda F1 were told that the car being readied for 2009 at would be a very good one. In which case why not run the car in the first few races? However, the decision to withdraw from F1 was taken. The end product of this a few months later was that Honda F1 would re-emerge as Brawn GP. Brawn GP was formed on 6 March 2009, when it was confirmed that Ross Brawn, the former Technical Director for the Honda Racing F1 team, Ferrari and Benetton teams, had bought the F1 team from Honda.

The Brawn GP car (BGP 001) was originally designed with the intention of becoming the Honda RA109. In fact, owing to the poor performance of the Honda team in 2008, the team made a relatively early start in designing the 2009 car. However, following Honda's withdrawal, development of the car continued, in the hope that the team would be somehow rescued. On its racing debut in the first Grand Prix of 2009 in Australia the team finished first and second. Jenson Button won six of the first seven races of the season and in October at the Brazilian Grand Prix, he won the 2009 Drivers' Championship. Brawn GP won the Constructors' Championship. The team won eight of the season's seventeen races, and by winning both titles in its only year of competition became the first to achieve a 100% championship success rate.

On 16 November 2009 it was confirmed that the team's engine supplier, Mercedes-Benz, had purchased a 75.1% stake in Brawn GP which was renamed Mercedes GP for the 2010 season. It may be very wishful thinking on my part but could it have been Honda F1 that won the F1 Constructors' Championship if it had persevered into 2009?

Back on 20th February 2009 Takagi-san made another announcement regarding the Honda organization. What he said had serious implications on Honda business in the UK and on HME staff. There would be some reduction in staff at HME and this was coupled with a decision to cut costs. In my case I was told to stop travel to Europe unless absolutely necessary.

My reaction to these business decisions was to say that my long experience in the motor industry taught me that in times of economic recession or tough trading periods you do not stop doing things, as in the case of the F1 team, car production at Swindon etc. Instead you all work smarter to survive through these economically difficult and tough times.

One thing I learned clearly from my time working for motor car companies with their Head Offices based in Germany (BMW) and Japan (Honda) was that all the big business decisions of these companies, for example, those affecting their global vehicle production operations, were made by their top management in their Head Offices. One such big decision was that affecting Honda's car production at Swindon in the UK. In February 2019 Honda announced that its Swindon plant would close in 2021. Closure resulted in the loss of about 3,500 jobs in the area. Not many months earlier Honda had said that it had no plans to close the Swindon plant!

The end product of the organizational discussions referred to by Takagi-san in February 2009 was that I retired early from Honda Motor Europe on 6th April 2009. When Takagi-san arrived as the new HME President I asked him what his plans were for HME. He told me he would like to leave in a few years' time with HME run by Europeans.

Maybe by the time I left HME it was partly run day-to-day by Europeans but I still believe all the big business decisions affecting Honda in Europe were and would continue to be made by Honda's Japanese Directors at its Head Office in Tokyo.

Chapter 20
# Consultancy 2009–2019

**ON 3RD APRIL 2009** I sent an email to all the top management in HME's Head Office, Honda's European Genpos and Honda's European independent distributors.

*Subject: Leaving Honda.*

*Dear colleagues,*
*This is to let you know I am leaving Honda Motor Europe on 6th April 2009, and I will be establishing an automotive consultancy which will offer a wide range of support services in Europe and internationally. From 7th April 2009 my email address will be john.sparrow@autoconsult.org.uk*
*I want to take this opportunity to say I have enjoyed working with you over the past years, and I wish you and your team success in the future. Please forward this email to other members of your team on a 'need to know' basis. Thank you.*
*Best regards, John Sparrow*

Fortunately having retired early from both BMW Group and now Honda Motor Europe with good BMW and Honda pensions I was not in any need financially to look for work. However, I did not want to stop working totally so I decided I would offer my considerable automotive knowledge and experience to all the main players in the motor industry, including motor manufacturers, automotive distributors and dealers.

I was very pleased that my email resulted in contact from a number of Honda's distributors enquiring about the kind of consultancy I could provide. I followed these up and this led to some good consultancy work being achieved via visits and meetings with many Honda independent distributors in Europe during May to July 2009. It was ironic that what I now provided was available to these distributors from me only a few months earlier free of charge. Now they were willing to pay me a reasonable fee for my advice and support to their businesses. I was delighted to offer my consultancy support to these Honda distributors shortly after leaving Honda. Later I was also able to offer my consultancy advice to two of the world's big four accounting firms.

Coincident with my communications with Honda's European distributors I discussed with Chris Rogers, who left HME the same day as me, how we could use our combined automotive knowledge and experience. Chris had been Head of Corporate Affairs for Honda Motor Europe with strategic responsibility for corporate public relations in the Europe/Middle East/Africa region and public affairs in Europe.

Chris and I met to consider our possible options. We agreed to pursue potential customers in our own areas of expertise. However, Chris said he believed there was a potential opportunity to create what he called 'a UK Automotive Forum' to improve communications in the wider UK automotive industry. We produced a presentation as a basis for discussion with key automotive industry personnel well known to Chris and/or me.

I could write a book on this subject alone. I have kept among my computer files the detailed notes I took of over thirty meetings Chris and I held in 2009 as we sought to communicate with and gain support to our UK Automotive Forum proposal from all sectors of and key stakeholders in the wider UK automotive industry. These notes include the positive and supportive words we received at all these many meetings.

These meetings included senior automotive executives, automotive suppliers, Conservative, Labour and Liberal Democrat MPs, one MEP and one Lord, all with a direct interest in the UK motor industry. In addition, we met senior executives with direct automotive interest from the government's Department for Business Innovation & Skills (DBIS), SMMT, global accountancy firms, seven Regional Development Agencies, and the Welsh Development Agency.

It is important to note here that the SMMT (the Society of Motor Manufacturers and Traders) is the trade association for the United Kingdom motor industry. Its role was and still is to 'promote the interests of the UK automotive industry at home and abroad' and it represents more than 800 automotive companies in the UK, including motor manufacturers and dealers.

Having met with the President and Chief Executive of the SMMT it was agreed by them that as the SMMT represented UK-based motor manufacturers, meetings with them was not necessary.

Everyone we met, including the SMMT, agreed in principle to our idea and proposal. The positive outcome of the discussions was that it was agreed there would be a first UK Automotive Forum conference on 15th June 2010 in the House of Commons.

However only a few weeks later I received a phone call from the Chief Executive of the SMMT confirming the SMMT could no longer support a UK Automotive Forum as proposed by Chris Rogers and me. This decision was confirmed to me in an email from him on 12th July. It said they did not think it was appropriate to support a further initiative. Our last communication to all stakeholders who had supported Chris Rogers and me said the UK Automotive Forum was being 'mothballed'. I knew it was very unlikely to be 'un-mothballed' in the future. My feeling at the time was it was a great pity the motor manufacturers were not briefed and informed at the outset of our journey in May 2009 by the SMMT, so Chris and I could have avoided all the time, effort and cost, as well as all the meetings where we received only encouragement by everyone we met on our UK Automotive Forum proposal.

Earlier in 2010 I was invited by Professor Peter Cooke of the University of Buckingham to chair a number of Automotive Forums held at the university. I had met Peter at some automotive conferences in Europe many years before. At the April event I met David Cardle, who was Managing Director of Frazer-Nash Associates, a consultancy specializing in automotive logistics. We then met to see how we could combine our extensive automotive knowledge and experience.

On 11th March 2011 an undersea earthquake of 9.0 magnitude occurred off the north-east coast of Japan. It was the fourth-most powerful in the world since records began and triggered powerful tsunami waves that reached heights of 40 metres. The tsunami swept onto the Japanese mainland, killing over 10,000 people. An estimated 230,000 cars and trucks were destroyed or damaged in the disaster.

The tsunami had a significant impact on the production and distribution of motor vehicles in Japan. I agreed to do some research and make a presentation at the 13th September 2011 Forum at Buckingham University. It was titled 'The implications of the Japanese earthquake and tsunami for the UK motor industry'. I will cover just a few facts here.

In addition to the serious impact in Japan there was also adverse impact on global vehicle production. Automakers across the world including the UK depended on Japan to deliver sophisticated specialized parts such as batteries, memory chips, dashboard displays, transmission controls and electronic components of the computer systems that power modern vehicles. One example was a shortage of Xirallic, a shiny pigment used in automobile paints. It was only made in Japan by Merck, a German chemical company in Onahama, a coastal town damaged by the tsunami and exposed to radiation from the damaged Fukushima nuclear power plant. The lack of Xirallic emerged as a big problem. Another subject to emerge from this disaster was 'single sourcing', that is, relying on just one source of supply for a component or product. Honda's response was to say, 'We will think of ways to change to dual sourcing over time.'

I carried out a number of automotive projects for Professor Cooke at Buckingham University. One example was the Royal Automobile Club (RAC), which had been bought by Carlyle from Aviva in June 2011. Now we were asked to carry out a project to measure and record the activities of RAC patrols. Briefing notes were produced for the patrols involved, starting with: 'The RAC has decided to carry out a study the purpose of which is to review current operational practice of RAC patrols and to identify best practice for customer handling processes.'

At the end of November 2012 I went out with several RAC patrols on their twelve-hour day shifts in Oxfordshire and Berkshire to observe and record what happened in their working day. My colleague John Foster carried out similar visits with RAC patrols in Yorkshire. I cannot reveal the main points emerging from this work except to say we produced a very detailed report for the RAC which included many recommendations to improve the RAC's operational efficiency and RAC member satisfaction.

I found it a very interesting experience as it was not something I had done before. One thing I remember was the relatively large number of RAC members we went to help who had tyre punctures. Many of these vehicles had no spare tyre, and, even worse, many owners didn't know their car had no spare tyre. As a result we could not help many RAC members on our visit as the RAC patrols did not carry spare tyres.

Nowadays car manufacturers have largely replaced the vehicle's spare tyre with a tyre repair kit. It's a way of saving money and reducing the weight of the vehicle to improve its fuel economy.

That theory is maybe fine if you have a small puncture and the tyre is largely still a tyre and you are not too far from home. However I have twice had major tyre blow-outs on motorways where there was no tyre left, and where a tyre repair kit would have been useless.

In agreement with John Spoerry, Head of Professional and Executive Development at the University of Buckingham Business School, David Cardle and I produced a training programme called 'A Comprehensive

Introduction to the Global Automotive Industry'. The programme followed a vehicle from design, product development, factory design and layout through manufacturing, logistics, distribution, retail, after-sales through to end of life and recycling.

I wanted to include the subjects of customer satisfaction and customer loyalty in the programme. This was a favourite subject of mine so I did some research. What I found was very interesting. Honda had among the highest levels of customer satisfaction in Europe but among the lowest levels of customer loyalty. Customers were satisfied with the quality and reliability but might not buy again. However, Land Rover in Europe was the opposite. It had among the lowest levels of customer satisfaction mainly because of quality issues but among the highest levels of loyalty. Customers were willing to put up with problems and buy again. That was also my experience of my many years working in Land Rover. As a business measure customer loyalty is much more important than customer satisfaction so I included a few slides on this subject in the training course.

At the end of 2015 we reached agreement with the Institute of the Motor Industry (IMI) to include a modified version of the training course on the IMI's Continuing Professional Development Programme (CPD). In February 2017, under my consultancy Autoconsult, David Cardle and I delivered a bespoke version of the two-day training programme for the IMI's staff at IMI's conference centre. Since then, we delivered a modified one-day training course for a number of companies including the Motor Ombudsman Ltd and the SMMT.

I started this book writing about my interest in motor cars from a very young age. So I decided I would end it by telling you the favourite cars I have driven in my lifetime. Number 1 is the Range Rover that I was fortunate to drive as a company car from 1983 to 1997 and to see its progressive development into a true luxury car. Number 2 is the Morris Minor, my first car from 1968 to 1970. Nothing much beats the enjoyment of owning and driving your first car. Number 3 is the Honda

CR-V, my car since 2010. Nothing much beats it for quality and reliability, important if you are a senior citizen.

I also decided in finishing my book to make a brief comment on the evolution of the UK car industry in those fifty years. When I started my career in 1969 BL was reputedly the fifth-largest motor manufacturer in the world. Now in the UK there is not one significant wholly owned British motor manufacturer. Austin, Morris, Riley, Rover Triumph, Wolesley don't exist as cars produced today. Jaguar and Land Rover are owned by the Indian Tata company. MG has been absorbed into the Chinese SAIC Motor. In 2019 the Italian investment group Investindustrial acquired a majority stake in the Morgan Motor Company, and in 2021 Japanese VT holdings acquired Caterham Cars.

I cannot imagine the Germans allowing even one of BMW, Mercedes or VW to disappear or fall into foreign ownership, or for the French and Italians to allow the same with their wholly owned car manufacturers. How did we allow that to happen in the UK? As I said before with all the big decisions being made at Head Offices outside the UK, will any of the current car factories of BMW, Nissan and Toyota eventually remain in the UK? I wouldn't put money on it. A subject worthy of a book itself?

Finally, what about the fuel to power cars in future? In the last fifty years the main fuels powering cars have been petrol and diesel. However in recent years car manufacturers have been required to lower the emissions of their cars. The UK government has proposed banning the sale of new petrol and diesel cars after 2030, thereby forcing those wanting a new car to buy an electric car. From 2035 the EU emissions target for cars is 0g $CO_2$/km although synthetic petrol will be allowed to power combustion engines. All this has meant a rapid move by car manufacturers to focus on the production of 100% electric vehicles (EVs). But EVs are not the ultimate answer. With their heavy batteries they are expensive to make and expensive to buy. Even in the future EVs will still be relatively expensive to make.

I am sure the fuel powering cars of the future is hydrogen, the most abundantly available element in the universe. Most car manufacturers know the future is hydrogen too. I will not be around to see it but even now what is called fuel cell technology is under development by many of the world's largest car companies. The Chairman of BMW AG recently said, 'Hydrogen ... has a key role to play in the energy transition process and therefore in climate protection.' A hydrogen fuel cell car emits only water vapour in the form of steam. So, zero emissions. This subject is certainly worthy of a book itself.

# Acknowledgements

I WOULD LIKE TO ACKNOWLEDGE the important contribution of the following people in helping and supporting me on the journey to publishing this book. I apologise if I have missed anyone who made even a small contribution – if I have, you know who you are.

Annette Sparrow, my lovely wife who gave me her love and support while I wrote the book.

Gillian Sparrow, my beautiful daughter, for encouraging me to write the book.

Adrienne Roche, a good friend of my wife Annette, and now me, for her support and help on my Findmypast research. This led me to find and meet Ros Flatt (née Bailey), whom I rented a house with in 1970, and is now a good friend who has given me her support. Ros kindly invited me to watch the coronation of King Charles III at her home, where I met her daughter Molly Flatt, an author and journalist, who recommended Whitefox Publishing Ltd to publish my book.

In the Whitefox Publishing team I acknowledge with gratitude and thanks the support, advice, patience and help of John Bond, Hannah Bickerton and Rosie Pearce in bringing my book to publication.

Thanks to Jim Obee, best man at my wedding, my good friend and expert photographer, for taking the photo of my Honda CR-V and for agreeing to its use on the book's front cover. Also for taking the photo of the first Land Rover and first Mini Minor on the cover flap, and not least for his support throughout the publishing process.

Brenda Strangeways, my sister, for her support during the publishing process.

Chris Rogers, my good friend, for taking the photo of me and allowing me to use it on the book cover.

Jeff Coope and Stephen Laing at the British Motor Museum at Gaydon. Stephen for providing words on the museum and for giving permission to use these and the image of the first Land Rover and the first Mini Minor on my book.

In this book I have written about my life and career in the car industry. To everyone interested in cars, I strongly recommend a visit to the British Motor Museum at Gaydon. The museum has very kindly provided some words and given me permission to put these on the cover of the book, and has also given me permission to include an image of the first Land Rover HUE and the first Mini Minor, two great British motor cars that are on display in the museum.

Milton Keynes UK
Ingram Content Group UK Ltd.
UKHW012152151123
432640UK00003B/20/J